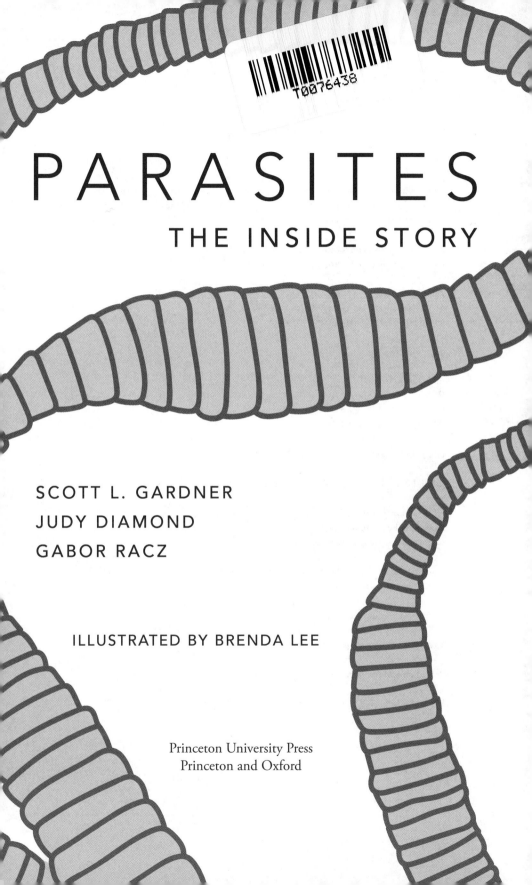

PARASITES
THE INSIDE STORY

SCOTT L. GARDNER
JUDY DIAMOND
GABOR RACZ

ILLUSTRATED BY BRENDA LEE

Princeton University Press
Princeton and Oxford

Published by Princeton University Press
41 William Street, Princeton, New Jersey 08540
99 Banbury Road, Oxford OX2 6JX

press.princeton.edu

ISBN 978-0-691-20687-5
ISBN (e-book) 978-0-691-24091-6

Library of Congress Control Number: 2022941988

British Library Cataloging-in-Publication Data is available

Editorial: Robert Kirk and Megan Mendonça
Production Editorial: Karen Carter
Cover and Text Design: Wanda España
Production: Steven Sears
Publicity: Kate Farquhar-Thomson and Sara Henning-Stout
Copyeditor: Lucinda Treadwell

Jacket/Cover Credit: Jacket art by Brenda Lee

This book has been composed in Adobe Garamond Pro

Printed on acid-free paper. ∞

Printed in the United States of America

10 9 8 7 6 5 4 3 2 1

PARASITES

THE INSIDE STORY

For Hadasa, Zsombor, Teal, Hudson, Grant, and Clark

Contents

Illustrations

Figures

Maps

Color Plates

These color plates follow page 110.

A Guide to Parasites Mentioned

Foreword

Human beings evolved as part of the global ecosystem—as one of the millions of species that it comprises—and we depend on it completely for our continued existence. Over the past five centuries, however, our numbers have grown from an estimated 500 million to nearly 8 billion. This unprecedented growth has turned our activities into an overpoweringly destructive force for the rest of life and the ecosystems themselves. Thus, we are already consuming some 70% more of the world's potential sustainable productivity than the total available (www.footprintnetwork.org). In the face of this relationship, we will be adding about 2 billion to our numbers over the next 30 years—by mid-century.

The global biosphere is incredibly complex, with probably tens of millions of evolutionary lines of bacteria interacting with perhaps 10 million or more species of complex cellular organisms—plants, animals, and fungi. We have given names to no more than a fifth of these, and even about the great majority of the ones we have named, we know next to nothing. Few people familiar with the situation believe that Earth can continuously support a human population as large as the one that exists now. In fact, we have already launched a great wave of extinction comparable to those that have occurred several times in the past. As much as a fifth of the plants, animals, and fungi that exist now may become extinct over the next several decades, and perhaps as many as twice that number, most of which have never been evaluated scientifically, by the end of this century.

We are not going to be able to learn everything about the world we are destroying as it goes, but we live in a time of unique opportunity to learn something about the variety of life here while it still exists. In this outstanding book, Scott Gardner, Judy Diamond, and Gabor Racz have given us a lovely picture of the importance and indeed fun in studying the members of one particular life form—parasites. Ultimately, it is the great satisfaction in learning about biodiversity that will inspire scientists to learn as much as we can while life is still relatively intact.

Parasites are unique and truly fascinating, as the pages of this book make so clear. They form close relationships with their hosts, the balance depending on both internal and external factors. Under one particular set of circumstances, the relationship will persist; under another, it may change, the parasites found in a particular host coming and going accordingly. To the extent that we can understand them, we are able to understand much about the stability and functioning of the hosts in which they occur.

Most extinction of species up to this point in time has been driven by human appropriation of natural lands for agriculture, including grazing, urban sprawl, and other uses. We have altered some 40% of global lands to produce food for humans, allowing villages, towns, and cities to grow and the elements of our civilization to be developed steadily in these centers. Now and in the years to come, however, global climate change is certain to become an even more important factor in driving extinction. We have already driven the average global climate to about 2 °F warmer than it was before we entered the industrial era some 150 years ago. Worse, national agreements to contain the rising temperatures have so far been seriously off the mark, and the average temperature increase may reach 2.7 °F (1.5 °C), said to be the point of no return, by the end of this decade, and move on toward as much as a 5 °F increase within a few more decades. Everyone is familiar with the disasters associated with warming temperatures—hurricanes, fires, rising sea levels—but the effects on extinction and the dismembering of ecosystems are even more profound. Much of our current agricultural land will become useless for the crops grown on it at present, with the numbers of starving people rising and climate-driven displacement of people widely evident.

We will be able to do the best possible job of creating a sustainable world if we base our decisions on as much information as possible. As the stories in this book tell so well, it can be a great deal of fun, as well as deeply important, to learn about the organisms that share this world with us. Parasites provide an especially sensitive and interesting key to the vitality of the systems in which they occur. They can also transfer to human beings, so that learning about them, as Scott Gardner, Judy Diamond, and Gabor Racz present so well in these pages, has a deep importance—they are endlessly interesting. I am sure you will enjoy and learn from these pages as much as I have, and I commend them highly to you.

—*Peter H. Raven*
President Emeritus
Missouri Botanical Garden, St. Louis

Introduction

Parasites are rarely described in positive terms. They are seen as blood suckers, freeloaders, scroungers, flunkies, deadbeats, and the worst kind of groupies. In Bong Joon-ho's 2019 award-winning film, the main characters first help the members of a wealthy family by tutoring the kids, cooking, housekeeping, and driving. Eventually the host family becomes dependent on the help, and only then does the relationship turn toxic, hence the film's name, *Parasite*.

In all natural and human impacted ecosystems on Earth, parasites are wildly abundant and represent a most successful lifestyle. They may live at the expense of their hosts, but both host and parasite are fundamentally changed as a result of the partnership. To understand how communities of organisms live together, one must know the parasites, since they play a central role in the dynamics of ecological associations. They are unseen influencers, affecting nearly every other species and contributing massively to the networks of interactions that stabilize ecosystems.

Dependent relationships between different species are the norm among living organisms, and these have evolved in every imaginable form. When the relationship between two different species benefits both, ecologists describe it as mutualism. Mutualism takes cooperation to a high degree of dependence as, for example, between tree roots and their mycorrhizal fungi, termites and their protistan ciliates harboring even smaller bacteria that digest wood, or lichens formed from a union of fungi and algae. When one partner benefits and the other is neither harmed nor helped, this relationship is referred to as commensalism. Examples of commensals are anemone fish that live sheltered among the stinging tentacles of sea anemones, tolerating their venom while being protected from predators, the tiny crabs that snuggle into the shells of oysters, or shrimp that live in glass sponges.

In theory, parasitism describes a long-term dependent relationship between different species where one benefits and the other is harmed. In practice, parasitism can range from deadly effects to cases where both parasite and host derive some benefits. The word *parasite* originally meant "next to food," and later the meaning shifted to mean sitting next to the host for free food. In most cases, parasites do depend on their hosts for essential nutrients, but how much harm the parasite causes is highly variable. In his classic book on ecology, Eugene Odum viewed parasitism in the same light as predation, but with the difference that parasites are more specialized in structure, metabolism, host specificity, and life history. Odum envisioned that the negative interactions of parasites in a community of organisms were balanced with positive ones in the form of commensalism and mutualism.

This book introduces readers to some of the challenging puzzles that confront parasitologists today. We first examine the human burden, the devastating toll that parasites have inflicted on human communities throughout the world. We then describe how parasites are scattered throughout the tree of life and focus in more detail on three of the most abundant kinds of parasites—the nematodes, the flatworms or platyhelminths, and the thorny-headed worms. These are all endoparasites—they live and thrive inside their hosts.

Finally, we explore the kinds of research conducted by parasitologists as they search remote locations for clues about the origin and diversity of parasite species. Scientists who study them are explorers in uncharted territory, since so little is known about the diversity, evolution, and ecology of these organisms that so dominate Earth. Parasitism is now, and always has been, a way to survive under changing environmental conditions. From arctic oceans to tropical forests, parasitologists investigate how parasites evolve, how they survive in changing circumstances, and how they influence the surrounding communities of organisms.

The Human Burden

Chapter One

Parasites on the Move

It was a long journey that lasted many thousands of generations. From clusters of settlements in what is now Asia, people arrived on foot and in boats, traveling in small family groups, settling for years, and then moving on. Some of these dauntless explorers crossed into North America via the Bering land bridge. During a worldwide glacial maximum period, the cool climate concentrated ice on land, resulting in lower sea levels and exposing a land bridge that made possible new routes for travel. With a resilient spirit, the early peoples left evidence of simple tools like spear points, skin scrapers, and hammerstones. And most certainly, they also brought along parasitic worms that had coexisted with them in their homelands.

The immigrants had no idea they were entering a continent undergoing massive faunal changes. Populations of mammals that had flourished earlier in North America were becoming rare. Within a few thousand years numerous species of large mammals went extinct. It was as if alien ships had scooped up most of the massive mammals from the planet, overlooking the few remnants. A more earthly process was undoubtedly at work—the relentless conversion of the landscape as the continent emerged from a glacial period. Plants on which grazing mammals had depended could no longer grow in the warmer, drier climate. Giant sloths, giant ancient armadillos called glyptodonts, and giant beavers gave way to puny descendants. Fourteen kinds of speedy pronghorn antelopes were survived by one remaining species. Camels, horses, tapirs, mastodons, and mammoths disappeared from their North American ranges. They survived in other parts of the world, but the

3

American bison remained as this continent's only large herbivore. As their sources of food disappeared, so too did their predators: American lions, saber-toothed cats, and dire wolves were replaced by smaller mountain lions, wolves, coyotes, and foxes.

The fossils of the original Pleistocene fauna are scattered throughout North and South America. These remains outline the story of the first peoples—what they ate, the kinds of tools they created, and how they buried their dead. Archaeological sites throughout the Americas also reveal evidence of the constellation of parasites that accompanied the migrants. Some, like *Enterobius* pinworms, were robust travelers and had no problem surviving in tropical or temperate climates. The incessant partnership between humans and pinworms goes back to a time before the common ancestor of humans and apes. Each human generation passed the parasite on to the next one—like DNA but not like DNA, since the transmission occurred not in the host cells but in the environment as the worm eggs moved among hosts. Pinworms are found in many kinds of primates, the group that includes monkeys and apes, but each species is remarkably specific. One close relative of the human pinworms infects chimpanzees, while others infect gorillas and orangutans. This suggests that different species of pinworms evolved parallel to the relationships of their hosts.

The pinworm, *Enterobius vermicularis*, causes one of the most common kinds of intestinal infection among people living in temperate zones such as North America. Their eggs are easily spread among children and people living in institutions via contaminated clothing, food, and surfaces, and they collect under fingernails and in bedding. Once ingested, the eggs hatch and the juveniles molt to adults, completing their life cycle in humans, their only host. Infection rarely results in serious illness. Because pinworms are relatively host specific, they act as a kind of marker that traces human movements over time. As humans migrated from one area to another, the pinworms came along, leaving traces in coprolites, the fossilized remains of human feces. Even small genetic

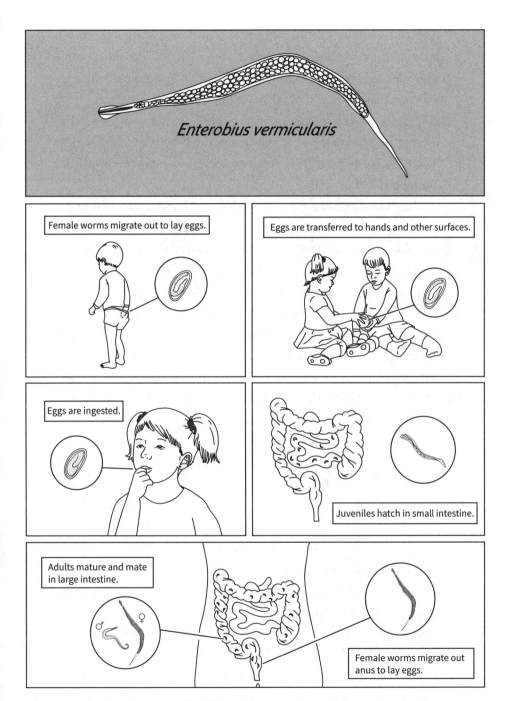

Figure 1. *Enterobius vermicularis* lifeline. Illustration by Brenda Lee.

differences within the same species of pinworm give clues to human movements. It turns out that the pinworms that accompanied people who migrated across the Bering land bridge to North America show genetic differences from those that accompanied other migrants. These differences aren't enough to consider the pinworms different species, but they do indicate that people traveled to the New World by multiple routes—some by land from Asia and some apparently by boat from Micronesia and beyond.

Many other parasites migrated with early humans from their origin in Africa to other regions, including the American tropics. Early archaeological sites in the Western Hemisphere reveal evidence of several parasites, including the large nematode *Ascaris lumbricoides* and the whipworm *Trichuris trichiura*. One species in particular—the human hookworm, *Ancylostoma duodenale*—is choosey about where it lives and reproduces, since its eggs and larvae don't tolerate a cold and dry climate. These worms probably accompanied ancient settlers who arrived in Latin America via other routes than the Bering land bridge, because it is unlikely the free-living stage of the juveniles could have survived through the Siberian cold.

Some parasite species are particular about having only humans as their definitive host. Others are opportunistic and settle for any large mammal. Of the 400 or so parasite species that infect people, most— as many as 70%—use humans only as incidental hosts. The fluke *Schistosoma mansoni* is an example of a parasite that uses humans incidentally, since it also readily infects chimpanzees, baboons, and rats. Dog and cat hookworms can be passed to humans—the tiny juveniles burrow into skin and cause infections—but fortunately the parasites don't reproduce inside people.

As people migrated, they carried parasites from their homelands, and during their travels, they picked up new ones. As they put down roots in larger and larger settlements, both kinds of parasites—those specific to human hosts and those acquired from other animals—were offered new opportunities for transmission. The settling of human populations into large stable aggregations allowed for more robust transmission of infectious diseases caused by viruses and bacteria, and it also enabled parasites to have more lasting and sometimes deadly impacts.

Human migrations have occurred throughout history. Sometimes invaders set out to distant lands to extract resources by conquering peoples, like the Spanish conquistadors in the sixteenth century, overpowering locals with their germs and weapons. But most people migrate because they are displaced—by war, by the collapse of their food supply, by infectious disease, and sometimes by racism and prejudice. The 300 years beginning in the 1600s mark an especially dark period in human history as more than 7 million people were forcibly sent from their homes in Africa to serve as slaves in the New World. The Portuguese, English, French, Spanish, Dutch, and Danish built slave economies throughout the Americas and other parts of the globe. People were captured from vast regions across Africa: first from what is now Senegal, Gambia, Angola, and Congo, and then from Togo, Benin, Nigeria, Mozambique, and Madagascar, and across the continent. From inhuman conditions on transport ships, slaves disembarked in the West Indies, Mexico, Colombia, and Brazil, and then were forced to work in forests, fields, mines, and homes.

Slave traders created inhuman conditions that enabled the movement of parasites with enslaved Africans. The parasitic protists that cause malaria, such as *Plasmodium falciparum*, probably originated in Africa, since they are closely related to those that infect other primates there. Throughout the course of the slave trade, different species and various lineages of *Plasmodium* from throughout Africa were introduced to the Americas. But Africans weren't the only source of the parasite. Analysis of DNA from at least three South American tribes suggests that there were earlier migrations to the New World from Australasia, since these tribes appear to share ancestry with Indigenous populations from Australia and Melanesia. One can imagine that at least one form of the malaria parasite could have arrived at South America before the European-driven slave trade.

The first enslaved Africans to arrive in the English colonies in Virginia may have been kidnapped from a Portuguese slave ship at around 1619. During the next years, hundreds of thousands of people from Africa were sold and traded to work on plantations. In the American colonies, enslaved Africans were highly prized in malaria-prone regions because they seemed to tolerate the disease better than Europeans. Some, in fact, carried a genetic mutation from their African homeland that limited the survival of the malaria-causing parasites by changing the configuration of hemoglobin in red blood cells. This mutation is called sickle cell disease, and it has been passed on through generations of survivors. The mutation still occurs in modern people living in areas where malaria no longer occurs; the mutation causes blood cells to reduce the amount of oxygen they carry, and under certain conditions the disease can be fatal.

Some parasites migrate with their hosts and then move comfortably from person to person in the new environment. Others require a suitable intermediate host in order to thrive. The intermediate host doesn't have to be the same species as that from the place of origin, since a close relative can serve as suitable replacement. The parasitic trematode flatworm, *Schistosoma mansoni*, probably first arrived in the Americas in people kept in the holds of slave ships. In Brazil, these flukes found acceptable intermediate hosts in local snails closely related to those in Africa. And the parasites flourished, continuing to infect people throughout the Americas wherever the snails and people lived together. The slave ships also brought the nematode, *Onchocerca volvulus*, that causes African river blindness. The presence of a blackfly host closely related to those in Africa enabled the parasite to naturalize in the New World.

Forced migrations continue to occur throughout the world, and the inhuman conditions of slave ships have been replaced by the squalor of refugee camps. In 2021, more than 80 million people were forcibly displaced from their homes. Two-thirds are currently from five countries: Syria, Afghanistan, South Sudan, Burma, and Somalia. When people from different regions are compelled to live in close quarters for long periods of time, their parasites are subjected to new intense selection pressures. By allowing the substandard sanitary conditions of refugee camps to persist, the international community empowers a witches' brew of microbes and parasites to spread well beyond the camps themselves.

When the first migrants came to the New World, they could not have known that gradual changes in climate had been reshaping the ecological landscape for millennia. The giant mammals that so dominated the Pleistocene landscape were in sharp decline. Over time, as these migrants became the first Americans, they adapted to live with the animals that remained, such as the vast herds of bison, that would sustain growing human populations in the plains. Colonialism changed all of that, first by subjugating Native peoples and then by decimating the herds of animals they had relied on. The colonists asserted their culture throughout the landscape, dominating the ecology with expansive human developments. Years later, this process has put into motion the conditions for much more rapid changes in global climate than the Earth has experienced throughout human history. And parasites are giving the first clues to how those changes will impact everyone on the planet.

Chapter Two

Parasites of Poverty

Even in the best of times, most people have one or more parasites living in or on them. One of them in particular is a gold medalist in the parasite Olympics, because it has managed to find its way into the guts of more than a billion people, making it the most prevalent parasitic infection of humans. *Ascaris* is an anomaly among parasites, since it requires no intermediate host, and it is nearly immortal: its eggs can stay alive stored for decades even in preservative.

Individual *Ascaris* are pencil-thin nematodes about the length of a spaghetti noodle. Put them all together and the biomass is staggering—inside a human, an adult female *Ascaris* produces about 200,000 eggs per day. The eggs are light and tiny—over the course of a year, egg production roughly equals the weight of two sugar cubes. This may seem like a small amount, but there are so many people infected with *Ascaris* worms that the total weight of eggs produced per year in infected humans worldwide is estimated at 66 million kilograms. This is a huge biomass, roughly equivalent to the weight of 350 adult blue whales, or 8,000 adult male elephants, or 360 fully loaded railroad coal cars.

Among living things, success is always a relative measure. For some organisms, like horseshoe crabs and the lobe-finned coelacanth fish, success equates with durability and stability over time, since these species have remained relatively unchanged for more than 400 million years. Most other species are continuously adapting in response to variable environmental conditions, and their survival can be the luck of the draw. The fossil record is littered with extinctions, not only of individual species but also of entire lineages. Parasites have a special problem in the struggle for survival, since their success depends not just on their own adaptations but also on those of their hosts. Choose a host on its way to extinction, and that could be the end of the parasite. The ability to switch hosts is a great insurance policy against ending up with a dud for a host. But host switching is a tricky affair with lots of uncertainty, and it always involves taking advantage of a random opportunity. From the parasite's point of view, it is a blind bettor's game, where the stakes are high and information about outcomes is virtually nonexistent.

Ascaris got lucky. Back in the age of dinosaurs, their ancestors' likely hosts were lumbering vegetarians called iguanodons, related to the duck-billed dinosaurs, and *Maiasaura*, the famous nesting dinosaur. Like modern-day sandhill cranes, the iguanodons migrated to nesting grounds to mate and lay their eggs. Parents may have warmed the eggs on nests by covering them with fermenting plant material, like a compost pile. The scientist George O. Poinar Jr. and his colleague Arthur Boucot, working in a site in Belgium, found ascarid eggs in coprolites from fossil iguanodons, suggesting the relationship between parasite and host goes back at least 125 million years in the early Cretaceous when these dinosaurs lived. At some critical point in time, some ascarids probably switched hosts to the pointy-nosed fur balls called multituberculates that fed mainly on insects at night, avoiding any unpleasant encounters with the fierce and agile small dinosaurs. Although these creatures eventually went extinct, their relatives gave rise to modern mammals. At least one of their parasite species, *Ascaris lumbricoides*, is the result of a lineage of nematodes that survived more than 100 million years to become the most common parasite of the most successful mammal on Earth.

The term *geohelminth* refers to nematodes that rely on soil contaminated with feces for transmission from host to host. Unlike tapeworms and many other kinds of parasites, geohelminths don't require an intermediate host. Female geohelminths can produce hundreds of thousands of eggs each day that are expelled in feces, and they depend on

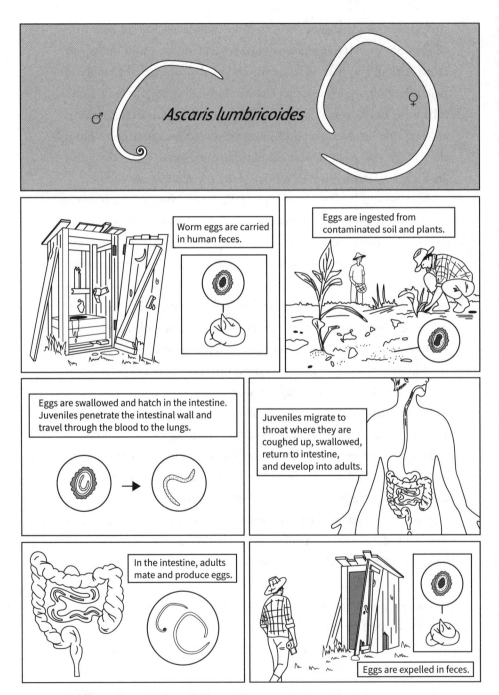

Figure 2. *Ascaris lumbricoides* lifeline. Illustration by Brenda Lee.

the likelihood that another host will somehow come in contact with the material. *Ascaris lumbricoides* survives by traveling from person to person via feces in relentless cycles of productivity. This reliable mode of transmission helped it earn its Olympic medal as a most durable internal parasite of people.

Humans infected with *Ascaris lumbricoides* can seem to have symptoms of flu that just won't go away. Once inside a person, the eggs hatch into juveniles that then make a circuitous trip from the intestine, into the blood, then to the liver, through the hepatic vein to the heart and the lung, then into the trachea and are then swallowed back into the intestine, where they become foot-long worms that mate and produce eggs. The presence of many worms in the intestine can cause abdominal cramps, nausea, fever, coughing, vomiting, eventual weight loss, and sometimes death. A sign of the close adaptive relationship between parasite and host is that *Ascaris* doesn't usually kill its hosts but continues to utilize them as active vehicles for spreading their eggs. Some estimates suggest that more than 50% of all children in the world are infected with *Ascaris* or other nematodes. They pick up the worms from playing in soil that contains fertilized eggs, from eating vegetables grown in fields that use human manure as a fertilizer, and from the myriad ways that kids come into contact with human waste when they lack adequate clean water.

Ascaris is only one type of nematode that causes problems for people. The phylum Nemata includes more species and more individuals of each species than any other major group of animals. There are more than a half million species of nematodes living in all kinds of organisms and in soils in tropical, desert, temperate, and polar regions. They are found in deep sea trenches and at the bottom of gold mines, and they are the most common animal on the floor of the oceans. Estimates of nematode abundance are staggering—on the order of 4.4×10^{20} inhabiting the upper layers of soil across the globe. That is a hard number to fathom. In 1977 the designers Charles and Ray Eames made a short film, called

Powers of Ten and the Relative Sizes of Things in the Universe, to help people get a sense of the meaning of really large numbers depicted logarithmically. By their estimate, if one considers the basic unit as a meter, then 10^{20} meters approaches the distance across the Milky Way galaxy. In other words, there are enough nematodes on Earth to line them up end to end and have nematodes in every meter across our entire galaxy. One can imagine that the scaffolding of our living world consists entirely of nematodes.

Geohelminths are relentless in their effectiveness at infecting humans and other animals. Three geohelminths—*Ascaris*, hookworm, and whipworm—form a formidable triad that together accounts for a major portion of worldwide parasitic disease. They are all nematodes transmitted through contaminated soil from person to person, and they all could be controlled by access to shoes, clean water, and adequate sanitation.

Another medal in the parasite Olympics might be awarded to the New World hookworm, a geohelminth nematode known as *Necator americanus*. The juveniles hatch from eggs in soil contaminated with human feces, and they feed on bacteria present there as they molt. Then they transform into mobile forms that infect people by penetrating the skin, preferably into the bare feet of a strolling human. Like *Ascaris*, they travel in the blood to the lungs, where they are coughed up and then swallowed so they end up in the intestine, where they molt several more times into adults, ready to mate and lay eggs. Infected people experience abdominal pain, weight loss, extreme tiredness, and anemia.

In the early part of the nineteenth century, when the United States was still a rural nation, most people lived in farm, mill, and mining communities. Hookworm was endemic in the farms and mining towns, particularly in the southern U.S. Few farmhouses and schools had access to an outhouse or privy. Although collection systems for human excrement were used by the ancient Romans as early as 2,500 years ago, sanitation in rural America remained a luxury for the elite.

In 1902, a zoologist working with the U.S. Department of Agriculture, Charles Wardell Stiles, presented a report on hookworm at a Sanitary Conference meeting in Washington, D.C. The resulting newspaper headlines announced the "Germ of Laziness Found" and captured the attention of a particularly wealthy individual, John D. Rockefeller. Seven years later, Rockefeller provided a million dollars to establish the Rockefeller Sanitary Commission for the Eradication of Hookworm Disease, which would attempt to map the prevalence of hookworm in the South, treat the 40% of the population who were infected, and eradicate the disease. The result was the first large-scale public health program established in the United States, and it was implemented despite the fierce resistance of the medical professionals who refused to believe that hookworm was the cause of the disease symptoms. The program created locally organized dispensaries that provided screening, medication, and public education about hookworm, and it led to the expansion of outdoor latrines throughout the southern U.S. Although the incidence of hookworm in the U.S. is now low because of improved sanitation, it remains a major source of disease for hundreds of millions of people living in developing countries with limited access to water and sanitary facilities.

Next in line for a nefarious championship in the parasite Olympics sadly goes to the third most common nematode to infect humans, the whipworm, *Trichuris trichiura*. Females are slightly longer than males, and they can produce tens of thousands of eggs each day. Once eggs are released via feces into soil and are ingested, they hatch in the host's intestine. Unique to this group of worms is a special organ that allows them to feed on the lining of the large intestine of their hosts, where they develop into adults and mate. About 700 million people, many of whom are children, are infected with whipworm in mostly tropical regions of the world. In children, a heavy infection has serious consequences, causing intestinal disorders, retarding growth, and impairing cognitive development.

The geohelminths have a long history of infecting people, and some scientists have suggested a coevolutionary relationship between these worms and humans. As people adapted to changing environmental conditions with advantageous immune modifications, the worms adapted in concert, developing better ways of damping their host's immune responses. In 1991 the well-preserved body of a 46-year-old Neolithic man emerged from a melting glacier in the Alps bordering Austria and Italy. News reports called him "the Ice Man," but he was soon given the name Ötzi, from the valley where he lived about 5,300 years ago. A flint arrowhead lodged in his back slit the artery under his collar bone and caused his death. But Ötzi was by no means a healthy man: he suffered from arthritis, and he showed clear evidence of infection by whipworm.

There is other archaeological evidence to support a long-term relationship between geohelminths and people. Whipworm and *Ascaris* eggs are known from the soils of archaeological sites in South Korea that date to about 4,000 years ago. Hookworm eggs containing developing juveniles have been identified from 2,300-year-old fecal samples from the Colorado Plateau. The oldest written reference to nematodes parasitic in humans was in the book *Huang Ti Nei Ching* or "The Yellow Emperor's Classic of Internal Medicine" from China about 4,700 years ago. Geohelminth eggs have also been identified in a 4,000-year-old Egyptian mummy and in even older fossilized feces from Brazil.

The majority of helminths that infect people are zoonotic, which means they occur in wild animals and infect humans opportunistically. The most prevalent human geohelminths—*Ascaris*, hookworm, and whipworm—have closely related species that infect other animals such as dogs, pigs, and nonhuman primates like baboons and macaques. Whipworm eggs have been found in 9,000 to 30,000-year-old fossils of the rock cavy, a medium-sized South American rodent related to guinea pigs, but these particular nematodes are specific to rodents, and there is no evidence of primates being infected by host switching from rodents. Recent evolutionary studies show that *Trichuris* species in humans and other primates share a common ancestor millions of years ago. Deforestation and loss of native habitats now regularly put wild animals in direct contact with people, and each time this occurs there is a potential for the transfer of both microbes and parasites. Sorting out the parasite family tree by fully understanding the evolutionary relationships among different species is an essential tool to managing present and future risks posed by parasites, particularly those as ubiquitous as the geohelminths.

Molecular studies have shown that geohelminths have profound abilities to manipulate and suppress their hosts' immune systems, enabling them to reduce inflammation and decrease the likelihood that the host can expel them. Studies on nematodes have led the science of genetics. *Ascaris lumbricoides* was the first animal to have its chromosomes studied, and a free-living nematode, *Caenorhabditis elegans*, was the first animal to have its genome sequenced. Subsequent genome studies on other nematodes have revealed the enormous diversity of their genes, and parasitic nematodes were found to be just as genetically diverse as free-living species. But the biggest surprise was that each species of parasitic helminth has independently evolved countless unique adaptations to its specific lifestyle, including its choice of hosts and the environments where it lives.

Chapter Three

Africa's Threatened Paradise

The Congo River is a majestic artery that nourishes one of the most diverse and distinctive regions in the world. The river is the second longest in Africa after the Nile, and its drainage basin covers a region larger than India. It is one of the world's deepest rivers, carrying vast amounts of water from highlands along the East African Rift, the fracture in the earth's surface where tectonic plates are ripping the continent apart. The Congo basin covers about one-seventh of African's landmass and provides food and fresh water to more than 75 million people from all walks of life—more than 150 different ethnic groups that include Bantu peoples, hunter-gatherers like the Ba'Aka, and immigrants from across Africa and beyond. The importance of the Congo basin to global biodiversity is hard to overstate. It encompasses a vital navigation and communications highway, and it is one of the world's largest potential sources of hydroelectric power.

Fish sustain the many people who live along the river and its many tributaries. Biologists are still sorting through the river's remarkable concentration of species richness, but so far about 800 kinds of fish are known. The waters abound with elephantfish with their trunklike noses and the ability to generate an electric field to find prey in the darkest water. The river houses more than 80 species of cichlids, the remarkably smart and social fishes that have elaborate parental care of their offspring and males that build nests and defend their young. The Congo has electric catfishes that can shock their prey with 300 volts, and the tributaries contain huge air-breathing lungfishes that can survive dry spells by burrowing into the mud. And beneath some of the most dangerous rapids in the world live tiny deepwater fish with shrunken eyes and no pigments on their skin.

At first glance, living alongside the rich biodiversity of the Congo River seems like a paradise. Easy access to abundant food and clean water is a luxury afforded to only the most privileged even in highly developed countries. In an alternative universe, the people of the Congo basin might have served as a world model for just and sustainable practices, but colonialism allowed no such opportunity. During the last half of the nineteenth century, King Leopold of Belgium laid claim to much of the Congo basin. He enslaved local people, greedily plundered resources for his own gain, and devastated long-standing culturally based communities, leaving social and environmental chaos whose scars remain to this day.

In the late 1920s a physician from Belgium, Jean Hissette, set up an ophthalmology clinic for villagers in what is now the Democratic Republic of the Congo (DRC). He found that in some river communities more than half the men over the age of 40 had impaired vision or had become blind from an unknown disease. Locals suspected a connection between living near the river and the catastrophic disease, and in an effort to avoid it, entire communities deserted the fertile river valleys for less productive uplands. In addition to blindness, the skin of those afflicted became scaly and itchy, with large nodules, and the disease became known as *kru kru* or *craw craw*. Hissette was the first to describe the source of the disease as infection by tiny parasitic worms called microfilariae. The disease would eventually become known as African river blindness or onchocerciasis, caused by hundreds of thousands of microfilariae in the lymphatic system of people. These tiny juvenile nematodes are released from adult female nematodes known as *Onchocerca volvulus* that live in the subcutaneous tissues of villagers. These little worms are one of the world's leading causes of blindness in people.

Onchocerca volvulus evolved as a human parasite. The only natural definitive hosts are people, and it spreads through vicious biting blackflies

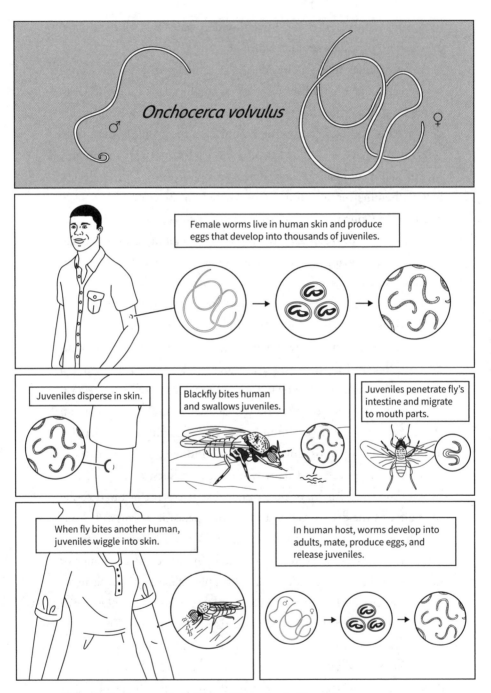

Figure 3. *Onchocerca volvulus* lifeline. Illustration by Brenda Lee.

of the genus *Simulium*, also called buffalo gnats. The flies live near fast-running streams throughout the Congo basin. Only female blackflies bite people; males feed on plant nectar. When a blackfly bites a person who has been infected with the nematode, it sucks in the tiny parasites along with the blood and lymph. The invaders then seek out the fly's flight muscles, where they develop through various juvenile stages, molting each time until they emerge as infective forms that migrate to the fly's salivary glands. When the female fly bites another person, the juvenile nematodes are injected into the human host, where they take up housekeeping in tissues just below the skin. There they develop into adults, mate, and produce eggs that hatch as microfilariae that spread through the lymphatic system of the person infected.

The threadlike microfilariae are thinner than a blood cell but are easily recognized in snips of skin examined with a microscope. Their long-term presence in a person triggers a disastrous immune reaction, mostly in response to the nematodes remaining in the skin. The inflammation leads to intense itching, skin thickening and cracking, and loss of pigment. Nodules just below the skin form around the adult worms, leaving infected people disfigured. When the microfilariae migrate to the eyes and die, the resulting inflammation scars the cornea and causes blindness. Worldwide, river blindness impairs the vision of a half million people and permanently blinds another quarter million.

Throughout history parasites have been a nemesis to humans, and there have been numerous attempts to control the diseases they cause. Some control programs focus on the vectors, attempting to reduce the availability of hosts as a means of killing the parasites. But those programs that rely on the applications of insecticides that have led some host species to develop resistance, so new, more potent insecticides are always in demand. The trade-off in evolution is that any control measure is effective for most, but not all, individuals in a large population. When those few survivors reproduce and repopulate, they spread resistance in subsequent generations. There is an ongoing tug-of-war between the efforts to develop chemicals that effectively kill insect hosts and the rapid ability of those hosts to develop resistance to the effects of these substances.

The effort to control African river blindness eventually generated an entirely original approach to treatment. Instead of trying to kill the intermediate host—the blackflies—it targeted the parasite in an unusual way. The approach was so remarkably novel that it led to a Nobel Prize in 2015 for the discovery of a new class of drug, the avermectins, and a derivative, ivermectin. The Prize was awarded to William C. Campbell and Satoshi Ōmura. Ōmura is a Japanese microbial chemist who searched for natural sources of medicines from the products produced by soil microorganisms. The scope of the task was beyond comprehension—soils can contain more than a billion microorganisms per gram. As Ōmura's soil samples were tested, about a third of them produced antimicrobial substances. One of the samples, gathered near a Tokyo golf course, contained a previously undescribed species of bacteria, *Streptomyces avermitilis*, which produced a variety of compounds called avermectins. Campbell then purified one of the active compounds in avermectin to produce the drug ivermectin. Ivermectin is truly a wonder drug that kills the juvenile stages of nematodes and parasitic arthropods such as lice, fleas, ticks, and mites.

Unfortunately, ivermectin kills only the developing stages of the river blindness worm, and the adult worms continue to live and reproduce inside infected people. Any long-term treatment requires repeated dosages for as long as ten to fifteen years. Recently scientists developed a new strategy that targets not only the worms, but also bacteria from the genus *Wolbachia* that live inside the river blindness nematodes. *Wolbachia* are common parasitic microbes found in more than half of all insect species, in other arthropods like mites and spiders, and in some nematodes, but is unknown from vertebrates. Like a set of matryoshka or Russian nesting dolls, the bacteria live inside the worms that live inside the host. Targeting the tiniest doll turned out to be a valuable step toward controlling one of the world's most devastating diseases. Ivermectin combined with an antibiotic works as treatment for people infected with river blindness. The ivermectin kills the juvenile stages of the nematode, and the antibiotic suppresses the *Wolbachia* inside, sterilizing the adult worms.

Ecologists often contrast the different kinds of symbioses as if they never overlap—relationships are either mutualistic where both partners benefit, they are commensals where one benefits and the other is neither harmed nor helped, or they are parasites. But as scientists learn more about the evolved relationships between organisms, the distinctions between the different kinds of symbiosis become wobbly, as if they don't quite hold up to what is observed in nature. Symbiosis can describe a continuum of effects ranging from beneficial to harmful to neutral depending on environmental conditions. *Wolbachia* might be the exemplar in this respect. One of the world's most universal microbes, it is sometimes described as an *obligate mutualistic endosymbiont*—an organism that can only live inside another and provides benefits to its partner. When *Wolbachia* living inside a nematode's cells are killed, the worm embryos can't develop, and the worm becomes infertile. This suggests that there are biochemical resources provided by *Wolbachia* that the worm cannot make for itself, thus indicating a clear case for mutualism. In other hosts, like fruit flies, infection with *Wolbachia* provides protection against RNA viruses.

Curiously, in other contexts, *Wolbachia* behaves somewhat like a parasite. In some insects the microbe circumvents the usual rules about the advantages of sexual reproduction: some varieties of *Wolbachia* kill males outright, others transform them into females, and still others induce female parthenogenesis, in which females develop from unfertilized eggs. Mosquito females infected experimentally with *Wolbachia* can successfully mate only with males that are also infected. Some populations of *Wolbachia* are known to have changed over time to become less harmful to their insect hosts, while others seem to maintain both harmful and helpful varieties across the overall population. The presence of *Wolbachia* triggers massive immune system responses that can be harmful to their mammalian hosts, and these too can change over time. In fact, many deleterious effects of African river blindness are not caused by the immune response to the worms but result from people's immune reactions to the *Wolbachia* as they are released when the microfilariae die.

Living with parasites is never just about finding the perfect medicine. In isolated villages throughout the Congo basin, access to treatment is a huge challenge. The most successful control programs involve local communities in prevention and treatment, along with support for those already infected. In the early 1990s, a young lecturer in public health at the University of Nsukka in Nigeria met a pregnant woman debilitated by itchy skin lesions and loss of skin pigment. The encounter led the lecturer, Uche Veronica Amazigo, to begin a life-long study of river blindness. Amazigo pioneered discoveries of the range of disabilities caused by river blindness skin disease, and she knew how to enlist the support of entire communities to combat the disease. She joined a women's support group and learned firsthand about the social effects of the disease on rural communities. She brought attention to the depth of social isolation, stigmatization, suffering, and disability caused by the disfigurement and unrelenting itching that results from infection, leading to global awareness about African river blindness and its social implications. Amazigo helped international organizations understand how to create community-directed treatment programs that would link research with organizational structures that empower communities to fully participate in distributing medicine for prevention and treatment.

The World Health Organization estimates that more than 20 million people are infected with the nematode worm *Onchocerca volvulus*, primarily in tropical regions of Africa, but also in some parts of Latin America and Yemen. Since there is still no vaccine or preventive medicine, control of the disease generally involves spraying rivers by helicopter or airplane with insecticide against the blackfly larvae and community-directed treatment with ivermectin and an antibiotic. To a large extent, these strategies are making progress: since 2013 four countries in Latin America—Colombia, Ecuador, Mexico, and Guatemala—have been declared free of the disease. The most isolated communities are often those most vulnerable. In the Amazon rainforest that straddles Venezuela and Brazil, more than 30,000 members of the Yanomami Tribe live in traditional villages, and river blindness is endemic in their communities. To combat the disease, health agents approached community elders and local shamans to seek their approval to provide tribe members with ivermectin. Yanomami are being trained to deliver river blindness treatment and education, walking jungle paths among villages to provide the medicine.

More than 300 species of worms infect humans. Some are rare or accidental, but about 90 are adapted to live in human hosts, and the diseases they cause throughout the world exceed the extent of both malaria and tuberculosis. The limited success of ivermectin has mobilized scientists to search for more effective medicines, and health organizations are more cognizant of the need to involve communities in treatment programs. It is easy to say that people have always lived with parasitic worms, and they always will. But the devastation to people's lives can't be underestimated, and huge challenges remain as to how to protect the culture and integrity of local communities, while also protecting biodiversity and controlling the most dangerous and destructive parasites.

It's a Beautiful Life

Chapter Four

Parasites in the Tree of Life

Parasitism is a lifestyle that shows up everywhere among living things, scattered like holiday ornaments on the tree of life. It has been said that every species of animal is either a parasite or a host. Among all known animals, there are more species that live as parasites than are free-living. Parasites are known to occur in nearly all types of organisms. Parasites have evolved in nearly every major branch of life and along many little twigs. They are found not just among animals since there are also parasitic plants and fungi. Viruses are parasitic by their very nature, and some—the phages—parasitize bacteria. Strikingly, there is only one major group of animals, the echinoderms, that are not known to have evolved a parasitic lifestyle.

Some of the parasites most harmful to humans are simple one-celled organisms that belong to a catch-all category called Protista. The scourge that causes malaria is a single-celled protist called *Plasmodium* that lives inside insects and people. Some plasmodia reproduce in the gut lining of *Anopheles* mosquitoes. When mosquitoes inject their saliva into blood, they transmit the tiny infective forms of the parasite to people. The developmental stages of some *Plasmodium* species are amazingly hardy and can remain in people for years, causing recurrent symptoms of malaria. Species of *Plasmodium* have lived with people for such a long time that parasites and humans have coevolved. People whose ancestors lived in West Africa can carry a gene that produces blood cells modified to keep the parasite out. But the presence of the gene can also produce sickle-cell disease, and the carriers, although more resistant to malaria, endure the deadly trade-off of blood that carries less oxygen.

Protists like *Plasmodium* seem to have made an evolutionary pact with certain insects that, in their role as hosts, enable transmission and wide distribution of the parasite. There are still more than 200 million cases of malaria worldwide, causing more than 400,000 deaths each year, the majority in children under the age of five.

A related single-celled parasite, *Toxoplasma gondii*, primarily infects cats, but it has far-reaching impacts on people. Exposure to cat droppings or eating undercooked meat have made this protist one of the most common parasites found in humans, infecting about a third of the world's population. Many symptoms of toxoplasmosis are mild, so many people aren't even aware they are infected, but this parasite is deadly for people with compromised immune systems, including babies, those who are HIV positive, and patients receiving cancer treatment. Recently, toxoplasmosis has been shown to be a major cause of mortality in sea otter populations in Northern California, where domestic cat feces with the resistant oocysts have been washing into the ocean and contaminating the food sources of the otters.

Another protist parasite causes African sleeping sickness. Tsetse flies introduce the protist *Trypanosoma brucei* into a person's blood, and the disease is caused when the parasite overwhelms the central nervous system. Although eradication efforts have dramatically reduced infections, there are still thousands of people in Sub-Saharan Africa infected with trypanosomes and thousands of others at risk of this disease. In the Americas, a related protist, *Trypanosoma cruzi*, is transmitted to people by reduviid bugs that transfer the infectious forms in their feces deposited on the skin of human hosts while the bug fills with blood. This trypanosome causes Chagas disease in millions of people in South and Central America, resulting in about 10,000 deaths annually. Still another group of parasitic protists includes the more than 20 species of *Leishmania* that infect people bitten by sand flies. The millions of people who are infected in the mostly tropical zones and warm deserts around the globe can endure terrible skin ulcers among their symptoms.

Aside from having only one cell, species of protists can be quite different from one another, and it is not clear whether they share a common ancestor. Most protists are free-living, but a parasitic lifestyle still features in thousands of known species. When it comes to exploiting another species' resources, being small has its advantages, and the single-celled protists are highly successful at the parasitic lifestyle. *Giardia duodenalis* is an unusual-looking protist parasite that seems to stare at an observer using a microscope with two eyelike nuclei that house its DNA. This protist opportunistically travels from animal to animal in water, and backpackers are often infected when they drink unpurified water. The reach of *Giardia* goes much further, since among people who lack access to clean water, up to one-third carry the parasite.

Parasitic lifestyles are well established among multicellular organisms. Many parasitic fungi infect plants, invading many of the important human food crops, like wheat, rice, corn, bananas, and especially potatoes. Their impact has had dramatic effects on human culture. In the 1840s infection of potatoes by a parasitic fungus led to mass starvation in a famine that created millions of refugees and forever changed the country of Ireland. The diseases that fungi cause sound like problems solved by a hardware store—rust, blast, blight, smut, and mildew—but fungi are so persistent and durable that sometimes the only means of protecting our agriculturally important food plants is by using naturally occurring strains of plants with fungal resistance. Modern agriculture, in which each individual food plant is genetically identical to the next one in the row, is highly vulnerable to the worst effects of parasitic fungi.

Most plants are free-living, but some species specialize as parasites on other plants, and these are strikingly diverse. The world's largest flower, the corpse lily, is part of a root parasite in the genus *Rafflesia*. So are pinedrops, the enigmatic little red or white lamp posts that pop up in forests as they parasitize a fungus that feeds on the roots of conifers. Pinedrops do not directly steal from the photosynthetic plant but rather from its fungal partner. The *Cuscata*, known commonly as dodder, are parasitic plants related to morning glory. Looking like a dump of old spaghetti noodles, they insert themselves into the vascular system of their host using a special organ called the haustorium. The European mistletoe, *Viscum album*, rather than living on the ground and feeding on roots, forms a clump high up in the canopy and inserts its haustoria into the tree's vascular tissues. Nearly every example of plant parasitism reflects an independent evolutionary event, evidence that parasitism arises repeatedly as a sustainable lifestyle.

The animal kingdom is less a realm than a heterogeneous collection of ways to be mobile, reactive, and multicellular. Among the 30 major phyla of invertebrates, there are huge differences in how often parasitism occurs. Some invertebrates, like the echinoderms, the phylum of animals that include sea stars, sand dollars, sea urchins, and sea cucumbers, have evolved no known parasitic forms. But another group, the sponges, shows evidence of parasitic lifestyles going back hundreds of millions of years. There are parasites among the jellyfish and corals, including *Myxobolus cerebralis*, which causes whirling disease in young salmon and trout, driving them to swim in circles while their nervous system is slowly destroyed. Among the segmented worms, there is a leech, *Placobdelloides jaegerskioeldi*, that lives its entire life feeding and reproducing in its strange habitat inside the anus of hippopotamuses.

Parasitism is common among the molluscs, the group that includes clams, mussels, squids, and snails. The larvae of all freshwater mussels and clams are parasitic on fish or tadpoles, and the adults have evolved ingenious ways to ensure that their larvae get into a proper host. For example, the fatmucket mussel, *Lampsilis siliquoidea*, attracts fish by waving a fleshy lure that mimics a tasty minnow. Once in close range, the mussel shoots out a cloud of larvae into the fish's gills, where the young mussels attach and feed until they are ready to live on their own.

With their skeletons on the outside and jointed limbs, arthropods are a remarkably diverse group that includes species of shrimp, crab, barnacle, centipede, isopod, flea, fly, bee, wasp, ant, tick, mite, and spider. Arthropod parasites have evolved some of the most unlikely habits for living in or on their hosts. For example, the shrimp *Typton carneus* lives inside fire sponges, using its claws to crush the rough sponge spicules so it can feed on the softer tissue. Some *Gasterophilus* bot flies lay their eggs on the hairs of the lower legs of horses, and the eggs hatch when the horse licks the eggs. After hatching, the parasitic larvae stick to the horse's tongue and migrate though the tissue to the stomach, where they finish their development feeding on blood and mucus. There are even parasitoid wasps that lay their eggs in the cocoons of other species of parasitic wasps.

One group of arthropods, the mites, has different species that parasitize nearly every other kind of animal or plant. Mites are a heterogeneous group that has affinities to spiders. Many are free-living in water or soil, and they only become a nuisance when a human strolls into their territory and gets bitten by their larvae, sometimes called chiggers. Some mites parasitize plants, causing galls on their growing tips. Others infect bees, and the *Varroa* mite has eliminated entire colonies and threatens crops that rely on the pollination services of the bees. In just one habitat—a pond—mites can parasitize nearly every other animal: one mite species infects damselflies, another lives on backswimmers, and others infect dragonflies, water birds, fish, and the occasional human child playing near the water.

Among all the organisms in the tree of life, there are three groups that excel in making parasitism a specialty. These three invertebrate groups include the nematodes, the flatworms or platyhelminths, and the thorny-headed worms. Nematodes are ridiculously common: estimates suggest that four out of every five animal species are nematodes. As many as 25,000 different species of nematodes are known to be parasites of vertebrates, and that probably represents only a fraction of nematode parasite diversity. Parasites abound in another huge group of worms—the soft-bodied flatworms or Platyhelminthes. About 80%

of all species in this phylum are parasites, making up the groups called flukes or trematodes, and the cestodes or tapeworms. And the third group is the most bizarre of all. Called thorny-headed worms, the Acanthocephala include almost 1,500 species—all of which are parasites—and, like medieval sorcerers, some exert powerful and strange effects on their hosts.

Animals related to humans—the vertebrates—are less united by complexity or culture, but rather by the utility and common origin of a steadfast structure, the backbone. Mammals, birds, reptiles, amphibians, and fishes make up the major groups of vertebrates, but among these, the parasitic lifestyle is relatively rare. Various vertebrates—some species of birds and fish, for example—are referred to as "brood parasites." Cowbirds, goldeneye ducks, red-headed ducks, and Old World cuckoos are species that lay their eggs in the nests of other birds, leaving it to the adoptive parents to raise the young. Sometimes the stealthy interlopers help the survival of their own young by throwing the other eggs out of the nest. In other species, the trespassing chick destroys the other eggs or kills the other fledglings, ensuring that the adoptive parents won't divide their attention and resources. Brood parasitism has evolved independently in different lineages of birds and fishes. This lifestyle is fundamentally distinct from the types of parasitism in which organisms, through evolution, change their body shape and functions in order to survive inside their hosts.

Parasitism is surprisingly common, and it is by no means limited to higher organisms. Early in the evolution of life, the mitochondria, the cellular organelles that provide energy to every higher organism, probably originated as parasites. Bacteria-like organisms adopted cells as their home base, eventually becoming an essential component of them. We don't think of them as parasites any longer because they currently exist only as integrated parts of living cells. Similarly, every virus depends on a host cell to reproduce, fundamentally a parasitic relationship.

All living things derive from common ancestors. The random mutations and recombinations of RNA and DNA generate the innovations that get tested through the trial and error of survival. Which individuals pass on their genes to future generations defines the tree of life. And along this sinuous path, among the many dead ends of extinction, there are a few vibrant trails that produce the organisms living today. Along nearly every one of those paths are organisms that have adopted parasitism as a means of survival. Usually smaller than their hosts, parasites live on the host's surface as *ectoparasites* or inside the host's body as *endoparasites*. This ingenious way of making a living has been adopted by all sorts of plants and animals, making parasitism the most successful lifestyle on Earth.

Chapter Five

A Perfect Host

Snails rarely attract attention. They tend to get noticed when they feast on garden plants or clean algae off the sides of an aquarium, but most of the time they hide under leaves, largely overlooked. But the parasitic flatworms known as trematodes or flukes have made snails the primary players in the distinctive stages of their lives. Snails are a trematode way station, an intermediate host that enables the young flukes to gain just enough independence to get to a definitive host where they can sexually reproduce. Parasitologist Armand Kuris, from the University of California at Santa Barbara, refers to hosts as islands—small patches within the vast matrix of habitat. For parasites, hosts are refugia that provide life sustaining resources during specific developmental stages.

Humans are among the few organisms that develop along a single continuous trajectory. Like most mammals, baby humans gradually grow into adults, changing the dimensions of their bodies, their intelligence, and their psychology, but otherwise remaining recognizable as one species. But for many, if not most, living things, abrupt transformation is the developmental rule. These animals have forms, usually called larvae, that are wholly unlike miniature adults. Frogs have free-swimming legless tadpoles that metamorphose into adults with four legs and sticky tongues. Butterflies have caterpillars, beetles have grubs, flies have maggots, crabs have nauplii, and clams and snails have veligers. In each case, an organism begins development in an entirely distinct form, and only later after metamorphosis—a complete cellular and tissue reorganization—does it become recognizable as an adult of the same species. Parasites, particularly parasitic worms, are champions at morphing from one body type to another. And often they transform more than once, each time changing body shape and physiology in predictable sequences during development.

Each stage has its own requirements, so that parasites need more than one kind of support system to make it through life. This means that many parasites require more than one host to survive long enough to reproduce and pass on their offspring. From a parasite's point of view, snails are an ideal go-between, providing nutrients and space to multiply before the parasite moves on to its next residency. Some parasitic flukes hatch from eggs deposited in water, and the young emerge as tiny swimming larvae with eyespots that enable them to detect light, a keen sense of smell to notice small concentrations of chemicals, and the ability to discriminate up from down. They move toward snails whose soft, slimy foot they can penetrate. In the snail the energetic larvae transform into little bags of wildly growing cells that multiply like crazy. The next step is when the bags break open, releasing new squirming offspring that may cycle again or blast out of the snail, swimming widely in the water to begin their search for their final or definitive host, like a turtle, fish, bird, dog, or person. Once they penetrate the skin of the new host and reach a suitable organ, like the liver or intestine, they transform into adults with male and female parts and begin to have sex with other adults of the same species. Soon new batches of eggs are passed into the water in the host's feces, hopefully landing in the water somewhere near a snail.

The neat pattern of movement from water to snail to definitive host is not universal among trematodes. Some species have the flexibility to vary the number of their hosts—whether it will be a snail and one other host, or with just the right conditions, two more. Eggs from the trematode *Coitocaecum parvum* hatch in water producing ciliated miracidia that swim in search of a snail. Inside the snail, each miracidium transforms into a sporocyst that reproduces asexually. These give rise to cercariae that exit the snail and swim out in search of their next host, an amphipod. Amphipods are small scavengers related to crabs with laterally compressed bodies that make them look like little books with feet sticking out from the pages and a head peering out with big eyes. Inside the amphipod, the parasite might remain for the rest of its life, producing eggs that are released when the amphipod dies. But if the amphipod host gets eaten by a fish, like the common bully *Gobiomorphus cotidianus* in New Zealand, the parasite undergoes another developmental leap. Inside this new fish host, the trematode begins to sexually reproduce, generating eggs that pass into the water.

Sometime in the remote past, snails and trematode flatworms formed an accidental association that has since become enduring, remarkable, and sometimes bizarre. Scientists know that the definitive hosts of trematodes, in which they develop into adults, mate, and sexually reproduce, are vertebrates like fish, birds, or mammals. But it is not known if snails started out as the sole hosts for trematodes, or whether snails were somehow acquired later. Studies of the evolutionary relationships among the different trematode species and their hosts will someday illuminate the origins of these relationships, but pieces of the puzzle are emerging from studies of the natural history of snails that begin to explain why snails are such fixtures in the life histories of trematodes.

One amazing characteristic of snails is that they have been around for a very long time. The ancestors of snails appeared with the earliest animals, more than 500 million years ago when, for the first time, sufficient levels of oxygen in the ocean triggered an explosive diversification of life. Snails were around when the first amphibians emerged onto land, and they would have risked being stomped on by dinosaurs. And unlike the dinosaurs, many species of snails survived the asteroid impact at the end of the Cretaceous, perhaps to become a significant source of food for the small animals that struggled to survive in the aftermath.

Snails are extremely diverse and highly flexible in their ecology, so they are found nearly everywhere on the planet from the Arctic to hydrothermal vents in the deepest abyssal plains of the ocean floor. Snails are gastropod molluscs, kin to clams, oysters, and squids. There are about a hundred thousand living gastropod species and tens of thousands of extinct fossil species. Snails differ from other gastropods, such as garden slugs and the elegant and colorful marine slugs called nudibranchs, by building their own external shell that completely encloses their body. The ability to build one's own home arose many times throughout evolution, so the term "snail" includes many different groups—some specialize on land, some in fresh water, and others are solely marine. The pulmonate snails breathe air with something like a lung, while other groups breathe with gills in water. Over time snails have diversified to such an extreme that one can find lung breathers living in fresh or marine water and gill breathers living on land. Some graze on algae with a toothlike file called a radula, while others are voracious carnivores, and still others are opportunistic omnivores feeding on almost anything.

One group of trematodes, called schistosomes, have been infecting snails and human ancestors for well over a million years. Early human relatives, like *Homo erectus*, would have been convenient hosts when they built camps near freshwater streams or rivers. When schistosome eggs hatch, the larvae that emerge have a special name, reserved only for this stage in flatworm development. Called miracidia, these tiny creatures have a range of special powers that would challenge any Marvel superhero. They have plates of hairlike cilia that enable them to swim. Their special homing abilities enable them to seek out just the right species of snail to serve as a host. Once they find a snail, the miracidia immediately drop their ciliated shields, and gland cells release enzymes to dissolve the surface of the foot, allowing the worm to penetrate. Once inside and near the point of entrance, each miracidium transforms into a mother sporocyst which then gives birth to many daughter sporocysts. The daughters immediately head off to other organs of the snail including the digestive gland. Eventually the daughters make many fork-tailed cercariae which blast out from the snail by the thousands into the surrounding water. With the help of eyespots, these tiny propellant arrows furiously wiggle to the surface, then slowly drop down through the water, only to swim to the surface again. They repeat this for about three days, and if they don't find a patch of human skin to infect, they die.

The lucky ones find an unfortunate human wading or swimming in the water. The cercariae attach onto a spot of skin with their suckers and immediately drop their tails in floppy piles. They then secrete enzymes that dissolve the skin so they can push inside between cells. Once having entered their new host, they move first to the liver, then to the heart, and finally pump through the blood vessels that lead to the large intestine. While still in a blood vessel, a male matures first, then the female seeks him out, sliding into a split in his body where she will produce eggs. Sex is no quick fix: the males and females remain attached in perpetual copula for their rest of the lives, which can be up to 40 years. Eventually some of the eggs work their way through the wall of the intestine into the lumen where they are passed with the feces. Other eggs are washed via the bloodstream into the liver, where they get stuck in tissue and trigger inflammation that eventually destroys the eggs but causes major disease in the host. If the infected person happens to defecate into a freshwater source, the cycle begins once again. In the short run, an infected person might feel like they have the flu, but long-term infections can cause learning disabilities, anemia, malnutrition, seizures, and multiple organ failure.

There are at least 20 species of schistosomes, and each has its preferences and quirks. Schistosomes are very particular about what species of snail they infect, and the distribution and habitat preferences of the snails—stagnant or slow-moving water or fast-flowing streams—determines the distribution of the trematodes. Humans have struggled with how to control schistosomes, and there are some remarkable successes. In the early part of the twentieth century, Japanese researchers had worked out the life cycle of *Schistosoma japonicum*, and they knew the danger posed to workers in rice fields. For a short time, a national

Figure 4. *Schistosoma mansoni* lifeline. Illustration by Brenda Lee.

campaign invited school children to collect the tiny snail hosts, called *Oncomelania*, for half yen per container. Later the more drastic steps of cementing irrigation canals, draining wetlands, applying snail poisons, and medically treating infected people finally led in 1994 to the eradication of schistosomiasis in Japan. Regardless of control efforts, throughout the world at least five species of schistosomes continue to cause more than 200,000 deaths per year and countless long-term disabilities, making this tiny flatworm one of the most prevalent human parasitic infections.

Not every eradication campaign turned out well. From 1950 into the 1980s, the Egyptian Ministry of Health launched massive campaigns to respond to the high rates of infection by *Schistosoma mansoni*. The approved treatment at the time was multiple injections with tartar emetic, although now the standard therapy is an oral drug. At the time there was little appreciation for what else could be transferred in blood, and disposable needles were not yet in common use. The misguided campaign resulted in population-wide infection with the hepatitis C virus. Although treatment for schistosome infections has been modernized, Egypt still has one of the world's highest rates of hepatitis C infection.

Trematodes are very small, with bodies ranging from less than a millimeter to a few centimeters in length. In some environments, certain species are so common their biomass is staggering. Ecologists use food webs to map the feeding relationships among organisms in environments like marshes or ponds, indicating which are producers, and which are primary consumers, secondary consumers, and tertiary consumers. A common scenario might feature algae, a snail, a minnow, and a frog as characters. Add pondweed, a water flea, a dragonfly, top predators like perch or herons, and a web begins to take shape. As scientists have begun to assess the numbers of trematodes that live in snails in marshes, their findings are transforming the understanding of how food webs represent connections in a community. The biomass of trematode parasites in a marsh turns out to be huge, sometimes more than that of all the birds that visit the marsh. It appears that depictions of common food webs are leaving out the elephant in the room, or most precisely, the massive numbers of wiggly parasitic worms in the scene. Scientists now claim that the health of an ecological community can be measured by the biomass of parasites. Instead of being a burden to bear, parasites act as mortar that stabilizes diverse communities of organisms and keeps them connected.

Parasitism always involves a dynamic relationship between the parasite and its hosts. The presence of the parasite inevitably causes harm to the host, but it is in the parasite's interest to keep the host alive, at least for the short term. Hosts are constantly evolving ways to minimize or ameliorate parasites' harmful effects. Although over time some hosts evolve more effective immune responses to a parasite, the two organisms evolve in concert with one another, so that one side's stronger immune response also promotes the evolution of the other's ability to avoid the response.

Sometimes the host finds inventive ways of surviving with its parasites. Inside a snail, some species of trematodes can inhibit the development of a snail's reproductive organs, functionally castrating the infected snail. One species of snail, *Biomphalaria glabrata*, compensates by accelerating its reproduction with a sudden burst of eggs before the trematode *Schistosoma mansoni* can do its damage. In some California salt marshes, more than half of the California horn snails, *Cerithideopsis californica*, are parasitized by a dozen or more species of trematodes that destroy their reproductive function. The snails respond to the onslaught by maturing earlier to reproduce before the parasite takes its toll. And the snails don't have just one kind of parasite to deal with. It is not uncommon for one snail to be infected with multiple trematode species. Any variations in the host that help it to survive to reproduce will be selected for in future generations. Host-parasite relationships are less a victory between contestants than a compromise in which both parties meet their survival needs to some degree.

Trematodes have evolved with horn snails for a long enough time to produce some remarkable adaptive patterns. In the most straightforward scenario, a trematode like *Euhaplorchis californiensis* infects a horn snail in a southern California salt marsh. The trematodes produce cercariae that leave the snail in search of a second intermediate host, a killifish. Killifish belong to a huge group of small fishes common in fresh or brackish water. Inside the killifish, the cercariae transform into meta-cercaria cysts. When a heron or egret eats a killifish, the trematodes are carried along, and in the bird—the final host—they mature into adults, mate, and produce eggs. This is a rather ordinary three-host pattern for a trematode: horn snail to killifish to heron. Along the way, however, some species of trematodes become something that is not ordinary at all.

Some trematodes can take up long-term residence in the horn snail's digestive gland, where they continue to reproduce. In this strategic location, the trematodes establish colonies with divisions of labor. It is not a voluntary system or even an enforced one, but as with ants, it is a genetically determined caste system. Some young trematodes are born to become larger individuals that will reproduce asexually and send out their cercariae to find vertebrate hosts and eventually establish new sexually reproducing colonies. Others are born small with large mouth parts, and they specialize in defense, attacking and ingesting any other trematodes that compete with their colony. The battles take place inside the winding digestive gland of the snail, where these soldiers deploy to the mid-regions where invading trematodes are most likely to concentrate. In colonies of closely related individuals, what works for the whole might not be the best solution for the individual. The forces of natural selection can sometimes drive a species toward relatively inhumane solutions because the overall reproductive fitness of the whole colony is improved. Inside the snail, the fine dynamics of trematode social behavior are intriguing, and their complexity is only just beginning to be studied in detail.

At first glance, worms and snails seem to have a most tenuous relationship. Snails are hardly a welcoming host, and trematode worms appear ungrateful guests, rushing off to more desirable hosts where they conduct their more important business of sexual reproduction. But the presence of intermediate hosts like snails has made these parasitic worms highly successful organisms. Ecological success occurred through simplicity, stability, and complex systems of developmental transformations from one form to another. Each stage in the parasite's life seeks out a different host to serve as the environment where the next stage can be nurtured. The system has specificity—each stage generally requires just the right species of host. And yet, when the ecological need arises, each stage is capable of endless variations. No salt marsh, no tide pool, no salt flat can be fully described without searching through the intricate relationships among the parasites and their hosts. Within each environment are complex communities driving the ecology of entire ecosystems, forming the scaffolding for all interactions among organisms.

Chapter Six

Hoberg's Tapeworms

Modern humans like to think they are the supreme life form on Earth. Our expanding population of 8 billion so dominates the planet that more than 95% of the land mass (excluding Antarctica) has been modified to suit our needs. That we are invincible seems so plausible—the great hunter transformed to the powerful industrialist and then to the clever technologist. Human insight and an uncanny ability to adapt to change make us well equipped to overcome any challenges, particularly those presented by the so-called lower forms of life.

And yet lower lifeforms have a way of persisting, providing a window that reveals the ancient and nuanced relationships between humans and other species. Consider the tapeworms, a group of flatworms that evolved sophisticated adaptations to living inside other animals. Over millions of years these parasites established their own form of supremacy. There are more than 20,000 species of tapeworms, and probably every species of vertebrate has served as a host for at least one kind of them. Tapeworms are among the most austere and efficient of organisms: the anterior end or scolex has been modified into a holdfast structure with suckers that enable them to attach to the gut of the host. Some species have hooks that can be released and reattached as the worm moves up or down the small intestine, like a kid going up a slide the wrong way and then letting go and sliding down backward.

The "tape" in tapeworm refers to its long flat body, made up of a chain of reproductive segments called proglottids. They have no gut and sport a ladder-like nervous system that enables a sparse repertoire of sensations. They can sense location, each other, the presence or absence of food, and annoyances like competition from other parasites. They are inventively self-contained and massively reproductive. They scatter into the environment thousands of little packages, each containing up to hundreds of thousands of eggs. On land, there is always a good chance that the tiny eggs or packets will be consumed by some unsuspecting animal in the act of munching grass. But in the ocean, tapeworm eggs get lost in the vastness of the marine environment, and they can be hard to find among the more numerous microorganisms, diatoms, and algae. The ingenuity of this system of dispersal lies in the swarm—overwhelming the odds that the eggs will be ignored and increasing the probability that at least some will enter another living animal.

Like perpetual tourists, tapeworms shuttle from one environment to another throughout their lives. Once ingested by an intermediate host, the eggs release the first infective stage, a tiny larval form with six hooks called a hexacanth. Using its tiny hooks, the hexacanth tears through the gut of the host to penetrate into the body cavity. This is the first stage in a long series of travels from animal to animal, until it finally reaches its definitive host.

Parasitism is not a single feature like wings or foot bones that is continuously modified throughout evolution. It is a dedicated lifestyle that can be adopted by almost any group of organisms. Parasitism has evolved independently in plants, fungi, and most animal groups. As organisms evolve from distant ancestors to present-day forms, some acquire a parasitic lifestyle. In groups like tapeworms, that lifestyle became so successful that it has lasted millions of years. During their long evolutionary history, tapeworms diversified and evolved differences that are sometimes subtle and sometimes extreme.

The features that distinguish each species of tapeworm are specialized adaptations: size varies in different species from tiny stubs no more than a millimeter in length to gigantic forms that infect blue whales, growing to 30 meters or more. Other differences between tapeworm species can be miniscule—differences in the number of hooks on the scolex or a few changed genes that are evident only through DNA sequencing. The genomes of tapeworms are relatively small, about one-third the size of those of their distant flatworm relatives, the blood flukes. Even though they have a relatively small number of genes, tapeworms have a kind of superpower, an inherent flexibility that resulted in the evolution of features that enable tapeworms to adapt to many different kinds of hosts.

Scientists piece together the genetic and morphological changes that each species of tapeworm acquired throughout their evolution. By reconstructing the changes, they create a phylogeny. Whereas a genealogical map traces a person's history back hundreds of years, a phylogeny traces entire assemblages of species back millions of years. Phylogenies map lineages of common descent based on small changes in genes or physical features. In this way, researchers have begun to unravel the long and convoluted history of tapeworms.

The former United States National Parasite Collection, currently housed in the Smithsonian's National Museum of Natural History, contains more than 20 million specimens, representing the world's largest collection of parasites. Parasites are so diverse that unrecognized species are discovered whenever a parasitologist explores a previously neglected host group, ventures into a new geographic region, or occasionally receives samples of infected animal tissue. Much like a seed bank, the national collection was intended to permanently archive specimens and information about every kind of parasite known. It has served as a great library of physical, genetic, and ecological information about these organisms, forming a primary source of reference for understanding the history of parasites, their unique ecological roles in linking landscapes to broader geographic regions, and their benefits and potential risks of disease for humans and other animals.

Eric Hoberg first arrived at the parasite collection in the early 1990s, later becoming chief curator when it was based at the Agricultural Research Service's facility in Beltsville, Maryland. Hoberg started his career studying the ecology of sea birds. But along the way, he began to collect the parasites inhabiting these birds, so he started to study their evolution and biogeography as well. His exploration of parasite diversity and evolution eventually extended to an assemblage of about 325 species of birds ranging from penguins to albatrosses to petrels and gulls. Most marine birds had never been systematically explored for parasites. On the birds' feathers and skin, he found ectoparasites like fleas and ticks, but inside was an amazing treasure trove of nematodes and flukes and a diverse collection of tapeworms. On several occasions he even found unusual parasites called pentastomes or tongue worms, organisms distantly related to barnacles.

Hoberg became a world's expert on tapeworms, with a focus on those that live in marine birds. Tapeworms that live in these birds have a specialized lifestyle determined, in part, by factors like the distance from their nesting sites to their foraging grounds at sea. Tapeworms have adapted to live in pelagic seabirds that spend their entire lives on the open ocean. The life cycles of these tapeworms are complex, sometimes with more than one intermediate host. For example, some find ways for their eggs to be gobbled up by small crustaceans which are then swallowed by fish that are eventually eaten by seabirds. The details of the life history of these parasites remain only vaguely understood.

Over the past century most biologists have supported the idea that parasites evolve in step with their hosts through a process called cospeciation. Cospeciation describes how one population of organisms changes over time in concert with another. If a host population becomes isolated, then the parasites on those hosts become isolated as well. If the isolated host population then undergoes genetic drift or selection and eventually forms a new species, the parasite might be expected to form a new species as well. When the phylogenetic history of the host species is mapped out, the parasite's history should form a rough reflection of it. It was originally thought that parasites do not often switch hosts, so it made perfect sense that one could trace the evolution of a parasite alongside that of its host.

In the 1980s when Hoberg started to map the relationships between the evolution of seabirds and those of their tapeworm hosts, he faced a conundrum. It was clear that the worms and the birds had not evolved in lockstep. The tapeworms turned out to have a much deeper evolutionary history than their avian hosts—these parasites were an ancient group that originated 300 million years ago, well before the dinosaurs. All birds evolved from small, feathered dinosaurs about 150 million years ago. Like old wine in new bottles, the parasites were vastly older than their seabird hosts. So what hosted the tapeworms at the outset, if there were no birds around during their initial radiation?

Three hundred million years ago, the continents were locked into one huge land mass called Pangaea. At this time, giant swamps were being transformed into forests, and reptiles, freed from their dependence on water to reproduce, were diversifying on land. At first, tapeworms started out in marine and aquatic environments as they infected aquatic predators—perhaps their larvae were eaten by an ancestor of the coelacanth, the remarkable lobe-finned fish that is still found today. Or tapeworm larvae may have been consumed by giant prehistoric sharks. Whatever the definitive host was, once eaten, some of the larvae managed to survive in the predator's gut. Over many generations, this lifestyle proved so advantageous that the relationship became obligatory. As reptiles diversified, tapeworms may have found hosts in the predatory mosasaurs and the long-necked plesiosaurs, marine reptiles that dominated the Mesozoic oceans.

Tapeworms may have first taken to the skies inside the many kinds of pterosaurs, reptiles that flew above the other giants. When the most recent asteroid impact acidified the oceans, suffocated the marine dinosaurs, and triggered the extinction of the pterosaurs, three-quarters of all animal species became extinct—but not the tapeworms. Some tapeworms apparently survived the catastrophe by switching hosts to a group of hardy survivors—the ancestors of modern seabirds. This was made clear when Hoberg compared the phylogeny of the tapeworms to that of their hosts. Rather than evolving in synchrony, as predicted by the model of cospeciation, tapeworms had forged a new host association with marine birds.

Hoberg's research focused on one particular group of sea birds called the Alcidae, a family that includes puffins, auklets, murrelets, guillemots, dovekies, and auks. With their black and white tuxedos, alcids are beautiful and entertaining subjects, rather like small penguins, though the two groups are not closely related. Like wind-up bath toys, they use their wings for propulsion under water and their feet for steering. They spend most of their time at sea feeding on schooling fish and tiny shrimp, called krill; they come to land only to breed. Distantly related to gulls, alcids are known in the fossil record since the Eocene, 35 million years ago, when the oceans were relatively warmer and were inhabited by some of the first whales. Most alcids are small birds, ranging from about 80 grams to just over a kilogram. But at 5 kg the largest member to survive to modern times was the great auk, which scoured the waters of the North Atlantic for fish and small invertebrates. Prized as a source of food, fish bait, down for pillows and beds, and decoration for clothing, the species was in sharp decline in Europe even in the 1600s. The last two great auks were killed off the coast of Iceland in 1844.

Most alcids live close to seacoasts in the Northern Hemisphere. They are particularly common around the Bering Sea, at the apex of the North Pacific, south of the narrow gap that separates the Americas from Eurasia. The relatively shallow Bering Sea is best known for the ephemeral corridors that provided a gateway for the dispersal and isolation of land organisms from one part of the globe to another. When the cold phases of the fluctuating climate caused sea levels to fall, the Bering Sea was spanned by land bridges that allowed elephants, lions, bears, and voles to make their way to the New World from Africa and Asia, while camels and horses headed in the opposite direction. Most famously, about 30,000 years ago, the Bering land bridge provided the pathway for the arrival of the first people in the Americas.

Hoberg realized that the climate-induced variations in sea level made the Bering Sea a special kind of evolutionary laboratory for the study of marine parasites and their hosts. Each time a cooling climate extended glaciers globally, sucked up fresh water and caused sea levels to drop, exposing the continental shelves, the Bering Sea became isolated from the North Pacific and Arctic oceans. This recurrent process happened more than twenty times, and each time, populations of marine animals became isolated, and each time the glaciers melted and sea levels rose, the animals dispersed more widely. These repeated expansions and isolations shaped the life histories of not only tapeworms and other parasites in alcids, but also the tapeworms that live in pinnipeds, seals, sea lions, and walruses.

Hoberg's research examines how the distant past has shaped present-day parasite-host relationships. As Hoberg looked at tapeworms in other hosts, he began to understand the origins of two of the most common tapeworms that infect humans—the pork tapeworm, *Taenia solium*, and the beef tapeworm, *Taenia saginata*. Today, humans are the definitive hosts of both *Taenia* tapeworms; *T. solium* uses pigs as the intermediate hosts, and *T. saginata* uses cattle as the intermediate host. Pigs pick up the *T. solium* eggs from grass or soil contaminated with infected human feces. The larvae hatch and move through the pig's gut to its muscles, where they develop into infective cysts. When a human eats uncooked pork, the cysts open and the larvae migrate to the gut. Thorough cooking will kill tapeworm larvae, so it is only raw or undercooked meat that presents the greatest opportunity for infection. Regardless, throughout the world, about 50 million people are infected by one of these two tapeworm species.

Infections of *T. solium* are insidious since they can be transmitted by consuming either larvae or eggs. Globally, about 50,000 people die

Figure 5. *Taenia saginata* lifeline. Illustration by Brenda Lee.

each year when the *T. solium* tapeworm larvae hatch and migrate through the body, forming larval cysts in multiple organs, including the brain, causing the disease cysticercosis. *T. solium* is particularly dangerous for people because humans can serve as both the intermediate and final hosts, and infected individuals can autoinfect themselves, ending up with massive numbers of larvae in their brain and other organs. Infection with *Taenia solium* is today one of the primary causes of late onset epilepsy in people. *Taenia saginata*, on the other hand, is far less pathogenic and is not normally known to encyst in organs of the human host, but the tapeworms can continue to live, grow, and reproduce in the gut for many years.

In the late 1990s, Hoberg and colleagues set out to analyze the evolutionary history of these two deadly tapeworms. It had long been assumed that domesticated pigs and cattle had originally been primary hosts for pork and beef tapeworms, and that their association with humans was related to the origins of animal agriculture. But Hoberg realized that *Taenia* had been around a lot longer than the domestication of pigs and cattle, so the tapeworms likely had wild animal hosts. The phylogenetic history of parasites reveals stories about primary hosts, but it can often tell us about the other hosts in the chain of infection. The puzzle includes not only tapeworms and their descendants, but also the other characters who were around at the time: the herbivores which, as they chewed on grass, accidentally swallowed the tapeworm eggs, the predators that hunted the herbivore hosts, and even the communities of scavengers that shared the kills. In the case of *Taenia*, one of those characters was probably an early hominin ancestor of modern humans.

Recent research suggests that more than 2 million years ago, an ancestor of the beef tapeworm *T. saginata* parasitized hyenas and wild dogs in Africa. A human ancestor like *Homo erectus* might have picked up the tapeworms as they scavenged on hyena kills or shared meat with hyenas and wild dogs at lion kills. Over time the tapeworms adapted to human hosts, who then accidentally transferred them to their cattle.

When hominids expanded out from Africa into Eurasia, similar events could have occurred with the pork tapeworm. *Homo erectus* scavenging on bear kills might have picked up the tapeworms, which eventually adopted humans as a primary host. Later the humans passed the worms on to their domesticated pigs. In these scenarios, instead of getting tapeworms from pigs and cattle, early humans acquired them from scavenging the kills of wild carnivores and much later transmitted the parasites to their farm animals.

The neat symmetry presented by the concept of cospeciation, where parasites and host evolve in lockstep with one another, isn't the primary driver in tapeworm evolution, and it may not be how most parasites evolve. Tapeworms aren't fuddy-duddies who stay with a bad idea long after it has been discredited. Instead, these parasites are continuously opportunistic, switching hosts over time as possibilities for new colonization become available. All parasites show ecological and evolutionary conservatism—at first seeming rigidly tied to their host species—but then they reveal an extraordinary capacity to exploit opportunities in the face of ecological change.

Much of the distribution of life on Earth was, and continues to be, configured by processes of episodic expansion and isolation that result from climate disruption. The studies by Hoberg and his colleagues have enriched the intertwined story of tapeworms, birds, and humans, showing that tracing the evolution of parasites can reveal new insights into the factors that govern Earth's biodiversity during times of accelerating change.

Chapter Seven

Whale of a Worm

The Israelite Jonah spent three days in the belly of a whale before being vomited out onto dry land. It couldn't have been easy. Most likely he would not have spent the time alone, since the sperm whale usually hosted many other guests. The first creatures that Jonah met in the whale's gut would have been tens of thousands of relatively small nematode worms. Later on, he might have come across the 30 meter tapeworm known as *Tetragonoporus calyptocephalus*, hiding out in the intestine. Jonah and the tapeworm would have had an interesting time: they had to endure a 1,000 meter dive to the ocean depths as their host searched for a tasty meal of fish or squid. And Jonah might have gasped while the whale held its breath for an hour and a half, collapsing its car-sized lungs into a flattened balloon. The tapeworm, on the other hand, would have been just fine, since it requires little oxygen. Jonah apparently managed to escape after three days, but the tapeworm likely lived out its lifetime inside the whale, traveling through a million miles of ocean. Tapeworms have no internal limits on their life span, so they live as long as their host manages to survive.

Could Jonah really have survived three days in the gut of a whale? Sperm whales have 40 to 50 teeth on their lower jaw, but generally none break through the surface of their upper jaw. They swallow other large prey, like sharks and fish, whole, so it is entirely possible that Jonah would make it into the gut. The survival challenge would be how to get through the four chambers of the whale's stomach with little oxygen while immersed in digestive enzymes. Tapeworms manage this feat by entering the whale as tiny larvae that don't grow until they pass through the gauntlet of the stomach. The larvae produce proteins that suppress digestive enzymes and block the host's local immune response, or they encase themselves into little cystlike bubbles. Others mimic the mucosal lining of the whale's gut, so the host recognizes the tapeworm larvae as just a bit more of itself, rather than as an intruder. All in all, Jonah would likely have fared better disguised as a young tapeworm.

Once it finally reaches the small intestine, the tapeworm's attachment organ or scolex hooks on so tightly that it isn't released even when the whale suddenly breaches straight out of the water, landing with force that would clear 10 Olympic-sized swimming pools. Throughout its life, the tapeworm grows by adding one section after another from the head end, getting longer and longer. It eventually has thousands of these segments, called *proglottids*, and like a line-up of thousands of tiny accordions, they periodically stretch and contract. Each proglottid is an independent hermaphrodite, named after the famous offspring of the gods Hermes and Aphrodite. Like the original Hermaphroditus, each proglottid contains both male and female reproductive organs, and each produces both sperm and eggs. Tapeworms are partial to sex, and they pretty much engage in any method that results in fertilized eggs, a practical solution when one lives inside another organism and is never quite sure when another of one's species will come along. When the opportunity avails itself, proglottids can reproduce sexually when two from different tapeworms mate and cross-fertilize with the other individual's proglottid, or when a single individual self-fertilizes by looping back on its self.

Eventually proglottids drop off the tapeworm at the far end and are expelled in the whale's feces. A whale tapeworm is incredibly fertile, and throughout its lifetime a single individual can make billions of eggs. In the vast ocean, even billions can seem like a small number, and tapeworms have to bank against tough odds. They hit the jackpot when an egg begins its improbable journey from one host to another, into the first intermediate host like zooplankton, then on to one or more fish,

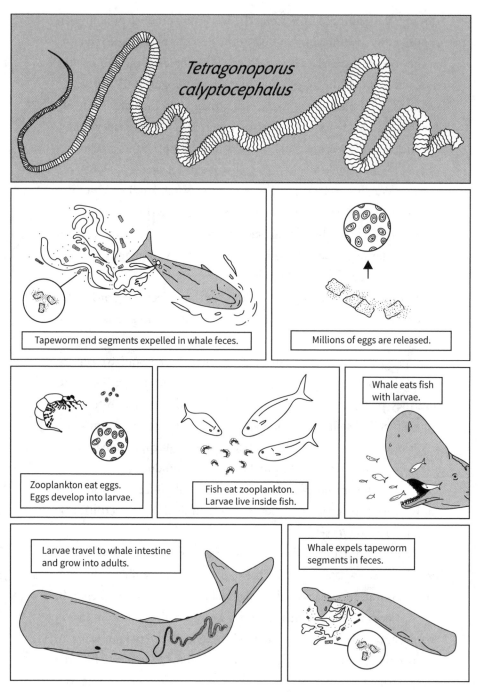

Figure 6. *Tetragonoporus calyptocephalus* lifeline. Illustration by Brenda Lee.

and finally into a whale, where the tapeworm becomes an adult that sends out more eggs.

There are 76 known living species of toothed whales that feed on fish and invertebrates and 15 species of baleen whales that feed on tiny animals, like krill. One might think that large size would have advantages for keeping other animals at bay, but that is not the case for these oversized mammals. Whales are magnets for parasites of all kinds. Barnacles clamp on by sucking the whale's skin into tiny tube-shaped cavities, and thousands of barnacles can homestead on a single whale, living for 20 to 30 years. The sea lamprey, a parasitic fish, fastens onto a whale with its suction-cup mouth while it rasps the skin and drains the host's blood. Whale lice aren't insects like other lice, but rather small crustaceans called amphipods that attach onto a whale's wounds and munch on its flaking skin. As if these parasites weren't trouble enough, kelp gulls eagerly grab flesh from the backs of southern right whales as they surface to breathe. The young right whales struggle to learn how to avoid the gulls by keeping their backs underwater. And that is just some of the eager fauna that whales have to reckon with on the outside of their massive bodies.

Inside the sperm whale's gut, Jonah would come across nematodes for a reason. They are probably the most common multicellular animals on Earth. About half of all nematode species are free-living, and the rest parasitize nearly every kind of animal or plant. And they are ridiculously common in the marine environment. In the 1970s, nematologist P. John Lambshead, from the Zoology Department at the Natural History Museum in London, set out to measure the species richness of marine nematodes—a count of the total number of species in that environment. Like a modern-day Alfred Russel Wallace, instead of searching for new species in the Malay Archipelago, Lambshead traveled across the North Pacific Ocean. He took core samples of bottom sediment at regular intervals to look for new kinds of nematodes, and in the process, he reshaped notions of marine biological diversity. His core samples averaged more than 100,000 nematodes *per square meter* even in deep-sea abyssal plains, a relatively impoverished habitat. Most astonishing of all, nearly every core sample had entirely different species of nematodes. Using these data, Lambshead estimated there could be as many as a million species of these worms on Earth.

In the 1950s a pregnant female sperm whale was found in the Kuril Islands, the archipelago that stretches from Japan to Russia. Living inside the placenta was the world's longest nematode, *Placentonema gigantissima*, at 8.4 meters. Imagine a parasitic worm as long as an adult giraffe is tall, with a thick outer cuticle, a few simple muscles, and a small mouth that leads to a long gut and anus. These worms have none of the specialized morphological characters associated with being an animal: they have no respiratory system, no vascular system, and few muscles in their digestive system to move food in and out. When conditions are difficult, some nematodes can suspend life processes to survive extremes of heat, cold, or desiccation, returning to the living when things get better. In their simplicity, nematodes have proven themselves resilient in nearly every habitat. Nematodes can even survive in the dry valleys of Antarctica and being dried and blown by winds to the highest level of the atmosphere.

Parasitic nematodes may seem invincible, but their struggle for survival lies in their ability to move from host to host, thus ensuring their development from egg to adult. Nematodes like *Anisakis simplex*, a common parasite of marine mammals, undertake a long journey through the oceanic food web. Their eggs might be eaten by tiny zooplankton like krill or copepods, where they hatch and encyst in the host. When the zooplankton are eaten by a fish or squid, the juvenile *Anisakis* burrow through the wall of the new host's gut and are encased in a protective bubble of tissue forming a cyst. This can happen multiple times as one fish with nematode cysts is eaten by another, and perhaps another. Each time an infected fish is eaten, the nematodes reactivate, move to the outside of the host's intestine, and then re-encyst. Eventually a whale or other marine mammal eats the massively infected fish. Once inside the gut of their definitive host, the worms leave their cysts, embed their anterior ends into the wall of the stomach, feed, grow, mate, and then release eggs into the ocean in the feces of the host. Sometimes baleen whales are directly infected when they consume zooplankton. One species, *Anisakis brevispiculata,* journeys through the marine ecosystem from zooplankton to bioluminescent lantern fish to dwarf or pygmy sperm whales. Humans are not normally hosts for *Anisakis*— they are instead a dead end for the worm—but when people eat raw or undercooked fish, they can have dangerous reactions. Some people have allergic responses and others get ulcers when the parasite embeds its head in their stomach. Sushi-lovers are commonly infected with this nematode.

Nematodes can be a mixed bag for marine mammals. The nematode *Crassicauda boopis* lives in various baleen whales, including fin whales. Fin whales are the second largest animals after the blue whale, and they can live to 140 years, but their adult nematodes can cause no end of trouble. With a shoelace-like shape, the nematodes can block the blood supply to the kidney, causing renal failure. Adult worms have been found in the kidneys of nursing fin whale calves, suggesting that the nematodes travel across the placenta or through milk to infect the young whales. But since none have yet been found in mother whales' placentas or mammary tissue, it is still a mystery how the calves are being infected.

Whales entered the oceans about 50 million years ago, when the Earth radiated with warmth and an overabundance of carbon dioxide, and the poles had little or no ice. Whales originated as land mammals, sharing ancestors with modern-day hippopotamuses. With their abundant life, the oceans must have been a rewarding place to feed. No one knows what drove ancestral whales to enter the oceans, but it was certainly a very gradual process, and only after millions of years did whales become so adapted to the marine environment that they could no longer survive on land. Some scientists have long believed that parasites evolve in tandem with their hosts, so one might expect to be able to trace the parasites of whales as they moved from land to sea. But either this perspective no longer holds, or something special happened during this time, because no parasites are known to have endured the transition of their hosts from land to ocean. None of the parasites found today in living whales have most recent common ancestors that also occur in or on land mammals.

Humans have hunted whales for centuries. Only in 1986 did the International Whaling Commission ban commercial whaling, and many countries continue to hunt whales under various loopholes. Whales continue to be killed by the actions of humans, whether from exposure to toxins and pollutants, strikes by ships, entanglement in fishing gear, disorientation due to underwater sonar, or starvation. Every year several thousand whales beach themselves on land, and most soon die. Many are toothed whales such as sperm whales, pilot whales, killer whales, harbor porpoises, or dolphins, but stranding of baleen whales does also occur. Inevitably, beached whales are found to harbor a rich fauna of parasites. The presence of parasites provides an easy explanation for the death of a whale. Regardless, parasites are abundant in both beached whales and healthy ones. Naturally occurring parasite infections could create pathologies in an injured or sick whale, but how they intersect with whale beaching is not yet understood.

Chapter Eight

Weaponized Worms

In the 1979 film *Alien*, an extraterrestrial infiltrates a spaceship deep in the cosmos, eventually laying its egg to hatch inside the body of the crew's executive officer. In the most memorable scene, the "chestburster" explodes out of the body of John Hurt's character, spewing blood and guts over the surviving crew. Heralded as one of Hollywood's scariest scenes, the creature was inspired by a 1944 painting by Francis Bacon, *Three Studies for Figures at the Base of a Crucifixion*. Hollywood films have made an art out of creating scary monsters, but they are poor matches for common creatures called acanthocephalans that occur naturally, living inside arthropod and vertebrate hosts.

Commonly called thorny-headed worms, these parasites are unlike any other worms. Once considered in a group all their own, recent molecular evidence suggests they share a common ancestor with rotifers, tiny ubiquitous aquatic animals that swim around and feed on algae using crowns of rotating cilia. But thorny-headed worms bear little obvious resemblance to their fluttery relatives. Some species reach as long as 65 cm, about the length of a tennis racket, and they have no digestive tract, absorbing nutrients directly through their body wall. Outdoing the best of scary aliens, they have an eversible proboscis with hooked spines used to pierce and grab onto the gut wall of their host. And that is only the beginning of the weird and awesome powers of this group of parasites.

Every one of the almost 1,500 known species of thorny-headed worms is a parasite. Their lives typically involve two hosts: an arthropod intermediate host where the young embryos develop, and a vertebrate definitive host. The intermediate host—typically a beetle, cockroach, or crustacean—ingests the worm eggs from contaminated soil, food, or water. The eggs hatch in the intestine, and the larvae wriggle through the intestinal wall into the body cavity, where they encyst and turn into the infective form. A suitable definitive host depends on the species of thorny-headed worm—it could be a rainbow trout, starling, raccoon, opossum, frog, lizard, whale, bird, or human. When the host eats an infected arthropod, the worm emerges, grows rapidly, drills its proboscis into the gut of the host, and proceeds to mature, mate, and produce eggs which pass out with the feces.

The 2018 French mystery series *Balthazar* features an episode in which a deranged young man synthesizes a powder called Dragon's Breath. When blown in the face of his female victims, they blindly follow his directives, in this case being told to commit suicide by hanging themselves from a bridge with a pink cord. The program could have been modeled on the natural and virulent abilities of thorny-headed worms to modify the behavior of their intermediate hosts.

Species of these worms have evolved a range of clever mechanisms by which they increase the odds that the infected intermediate host will be eaten, thus providing the worm with the opportunity to move on into its definitive host. Cockroaches and hedgehogs, for example, are no match for the devious mind control by the thorny-headed worm *Moniliformis moniliformis*. Infected cockroaches move only very slowly, so they make easy prey for hungry hedgehogs. Even if only a few roaches are infected, they are most likely to be eaten, so the parasite easily makes it way to its hedgehog host. Another species, *Pseudocorynosoma constrictum*, causes its intermediate host to look as though it is wearing a fluorescent orange traffic vest, perhaps making it more obvious to visual predators. The amphipod host is normally clear and inconspicuous, but when the larvae enter the host's body, the parasite turns fluorescent orange and shines like a lantern, possibly making it more likely that it will be eaten by a duck or other predator.

Rendering the intermediate host more obvious or slowing down its movements are only two of the ways these parasites increase their ability to travel to the next host. In ponds, lakes, and streams throughout much of Europe and North America, a small aquatic crustacean is a ready food source for birds, fish, and even some insects. Called *Gammarus lacustris*, this little amphipod is susceptible to infection by the thorny-headed worm *Polymorphus minutus*. The impact of infection is multifaceted: the amphipods get paler in color, and they begin to seek light and resist gravity, spending more time at the top of the water column, where they are more likely to be preyed upon by ducks and other birds. Scientists asked whether these responses might be specific to just this native species of amphipod, but it turns out that the parasite is just as effective at changing the behavior of a related amphipod, one that is a recent colonizer and considered an invasive species.

Parasitologist Janice Moore, from Colorado State University, wanted to know how effective acanthocephalans were at changing the behavior of their hosts. In the laboratory and in the field, she examined the thorny-headed worm *Plagiorhynchus cylindraceus* along with its intermediate hosts (isopods) and definitive bird hosts (starlings). She collected isopods from sites around Albuquerque, New Mexico, placed them in beakers with pieces of carrot covered with worm eggs, and then waited months for the parasites to develop in their intermediate hosts. Uninfected isopods were raised under similar conditions to serve as controls. Moore assessed the movement of the isopods under various conditions of humidity, shelter, and substrate color and found that the infected individuals all differed from the controls in ways that would make them more vulnerable to a bird predator. Her field tests with starlings then confirmed that the predators actually ate more of the infected isopods than they did of the controls.

Thorny-headed worms aren't the only parasites that manipulate their hosts. Imagine an infection that causes a person to climb up a very tall pole to be eaten by a passing Godzilla. And if Godzilla happens not to

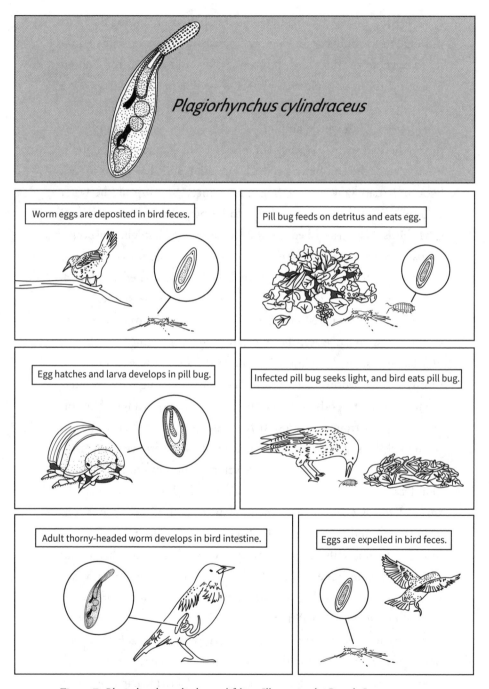

Figure 7. *Plagiorhynchus cylindraceus* lifeline. Illustration by Brenda Lee.

come, the person has to crawl back up every evening until the monster arrives. The fluke *Dicrocoelium dendriticum* is a parasite of many species of mammals in North America and Asia. It uses land snails and ants as intermediate hosts. Once an egg passes out of the definitive host in the feces and is eaten by a snail, it hatches as a miracidium that eventually transforms into thousands of cercariae. As the cercariae exit, they irritate the snail's mantle cavity, stimulating it to produce extra mucus that forms into balls of slime packed with the parasite. It turns out that the slime balls are a tasty treat for carpenter and other ants. Once inside an ant, some of the parasites move to the abdomen and thorax while others encyst in the brain, turning it into a nightly automaton. When the temperature goes down in the evening, the ant climbs up a piece of grass and holds on at the very top until morning. If it survives, the ant returns to its nest and goes about its normal activities until the next evening when it climbs back up a blade of grass. This behavior repeats every night until an unsuspecting deer or other mammal happens to chomp on that blade of grass, allowing the worm to infect its definitive host.

There seems to be no end to the ways a parasite can make its host look foolish. When the fluke *Leucochloridium variae* infects a snail, the cercariae climb up into the host's eye stalks, puffing them up like fat little caterpillars. The presence of the parasite makes the eye stalks pulsate, as if dancing to disco music, and it drives the snail to move into sunlight, where it attracts the attention of passing bird predators. Another fluke, *Uvulifer ambloplitis*, does its best to make its host look eye-catching. When it infects killifish, the cercariae penetrate the fish's skin and encyst there, and in response, the fish secretes melanin, which creates black spots. Infected fish with black spots are possibly more easily captured by kingfishers, the parasite's definitive host.

One famous case of host manipulation really does seem like a case of Dragon's Breath. Horsehair worms are classified into a phylum all their own. The free-living adults look like long strands of embroidery thread; they can reach up to 2 meters in length but are less than a millimeter wide. *Paragordius tricuspidatus* adults form crazy mating frenzies in fresh water that have been called Gordian knots, since they are reminiscent of the famous knot puzzle given by an oracle to Alexander the Great 2300 years ago. The larva of this species is a great manipulator. The eggs are eaten by snails and then develop into larvae which penetrate all of the snail's tissues. When the water level declines and the snails die, crickets come to feed on the dried carcasses, like kids eating popcorn at a movie. Once inside a cricket, the larvae invade the abdomen, develop into adults, and proceed to alter the expression of genes in the insect's nervous system. The cricket now begins to seek out water at night, possibly attracted to the reflections, where it is compelled to jump in and usually start to drown. The worms then emerge, return to their aquatic life, and begin to mate in a frenzy.

The interaction between parasite and host is often portrayed as an arms race, where one side tries to get the upper hand, the other retaliates, and so on until a victor emerges. And it would seem that the victor is often the parasite, which manipulates its host into submission and sometimes death. This is a very Hollywood conception of how organisms form relationships. By definition, parasitism is a type of evolved symbiosis in which one organism causes harm to another, but things aren't always so clear cut. More than half a century ago, the evolutionary biologist Richard Dawkins presented the concept of *extended phenotypes*, and parasites featured centrally in this idea. Biologists define the phenotype as an organism's expression of its genetic make-up—its form, morphology, physiology, and behavior—and it is the phenotype that is affected by natural selection. Environmental conditions act on the phenotype to influence survival, and when new genetic variations occur, they change the phenotype, potentially enhancing survival of the organism. Dawkins' idea was that the genetic changes in cohabiting organisms have mutual effects on each other, so their expression in the phenotype is actually a reflection of the genetics of both parasite and host. In this way, the genes of one member of the relationship directly impact the phenotype of the other member, through either physical interaction or alteration of gene expression. Natural selection acts on the expressed phenotype of the genetic variations of both host and parasite simultaneously.

Scientists have asked whether parasite-host interactions lead to beneficial ecosystem outcomes. Ecologists know that the presence of predators leads to healthier prey populations. When wolves were hunted nearly to extinction in North America, moose, elk, and deer populations initially exploded but were followed by huge die-offs, as the overabundant prey animals depleted their food. Reintroduction of wolves has stabilized many ecosystems, leading to healthier populations of both predators and prey. The same processes could be at work with parasites and their hosts, where the presence of parasites leads to a better fit between host populations and the available resources in their environment. For example, if kingfishers preferentially prey on infected killifish because their spots make them more conspicuous, perhaps uninfected fish that don't have spots are more likely to survive and breed. The critical issue is not only what helps each individual to produce more offspring, but also how the entire local population benefits. If infected hosts attract greater attention from predators, this might make the environment safer for hosts that haven't been infected. These examples suggest that a broader, ecological perspective helps to clarify the dynamics of host-parasite interactions. At first glance, it appears that parasites have the upper hand, but a closer look that takes into consideration the life histories of both parasite and host reveals a more nuanced story.

Expeditions

Chapter Nine

Hunting for the Origins
of a Deadly Virus

In 1993 people started showing up in tribal health centers in New Mexico with high fever and pneumonia-like symptoms that did not respond to antibiotic treatment, and many of the early patients died within a few days. The alarmed health care workers contacted state epidemiologists who brought in experts from the Centers for Disease Control and Prevention in Atlanta. Epidemiologists immediately began to look for patterns in how the disease spread through populations in the Southwest. In the case of the New Mexico disease, besides a hot, dry climate, the first common factor was the presence of large numbers of deer mice. Health alerts were soon sent out throughout the Four Corners region of the U.S. warning people to avoid areas with high concentrations of mice, particularly in sheds, unused barns, and cabins that had been closed for the winter. Called Sin Nombre or the "nameless," the pathogen, an *Orthohantavirus*, created widespread panic in New Mexico. It would turn out to be a virus called hantavirus that deer mice habored without ill effect but proved deadly when it jumped to humans.

This virus remains as deadly today as it was in 1993. In 2004, Jeff Kaminski, a graduate student from Virginia Tech, collected mice for his forest management research project in West Virginia. He died from a hantavirus infection after driving to his lab with traps full of mice stacked inside his car. There continue to be several known strains of hantavirus currently present in North America.

When the cause of this new disease was identified as a virus, scientists began to search for clues about its origins. It turns out that a related virus had been isolated in the early 1950s from Korean War soldiers in the Hantan River area—hence the original name "hantanvirus." That meant that a virus once present in Asia may have shown up in mice in the United States. Solving the puzzle of how the virus spread across the globe required that scientists take a close look at their hosts. Was the virus carried by global transportation systems, through the pet trade, or inadvertently through packages sent in the mail? Or had the virus always been associated with mice in the United States but had never emerged as an obvious human threat?

The researchers knew that the hantavirus identified from American soldiers in Korea had originated from a mouse similar to those found in Mongolia. Eurasian field mice in Mongolia are distant cousins of the common deer mouse, *Peromyscus maniculatus*. The deer mouse has the largest range of any North American rodent, occurring in a wide band from as far south as Oaxaca, Mexico, to as far north as the Yukon territories, and as far east as Labrador. The two species of mice from opposite sides of the globe are related through common ancestors that lived about 20 million years ago. At that time the climate was cooling, and as glaciers formed and hoarded water, global sea levels dropped, revealing new land bridges. The ancestral mice followed these land bridges into new environments, where over time, they diversified into new species. As the mice evolved, so too did the miniature ecosystems they carried with them—the intestinal bacteria, the helminths, the ectoparasites, and even the viruses.

At the time of the hantavirus outbreak in New Mexico, Mongolian researchers had been in contact with the New Mexico scientists because they shared an interest in setting up a long-term ecological research site. With support from the National Science Foundation, the group saw an opportunity to form an international collaboration to investigate the origins of the disease that had been ravaging people in the southwestern United States. The scientists suspected that hantavirus might have been present in the ancestors of both the Mongolian mice as well as those in North America.

Mongolia, located between two of the most powerful nations in the world—China and Russia—is the real Wild West. Mongolia contains the largest area of unfenced grasslands in the world, and it has the lowest density of people in any region of Asia. Many of the inhabitants of these grasslands are still nomads, herding their animals from winter to summer ranges following the seasonal growth of grasses. Mongolia's grasslands extend across the entire country from the Gobi Desert in the south to the forested taiga in the north. About twice the size of Texas, this expansive region encompasses all major biomes and associated ecosystems of Eurasia.

Mongolia is a massively tilted plateau, dropping from the west where the Khüiten Peak towers at 4,356 meters all the way down to Hoh Lake in the east at just over 500 meters. Covering more than 80% of the country, the grasslands of Mongolia are the largest wildlife habitat of its kind in Asia. Just the eastern portion is more than 10 times the size of the Serengeti in Africa, and in earlier times, its abundance of large animals was just as remarkable. Huge two-humped wild Bactrian camels, unique among mammals in their ability to survive on water saltier than the ocean, lived in small herds. Some were domesticated 5,000 years ago by early traders—these domestic Bactrians are now recognized as a separate species—but small herds of the original wild animals are still found in southwestern Mongolia. Now critically endangered, their future in the wild is uncertain. Mongolian saiga antelope, with their comical expressions and colorful ribbed horns, once blanketed the grasslands as they migrated in herds of more than 1,000. Today, they too are critically endangered: only a small population of 4,000 persists in far Western Mongolia. A similar fate awaits most of the other large native grassland inhabitants of Asia. Siberian ibex, Asiatic wild ass, and black-tailed gazelles are all threatened or endangered. The government

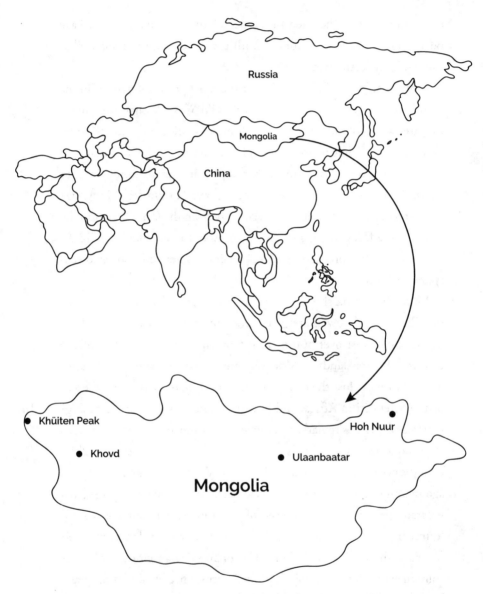

Map 1. Map of Mongolia. Illustration by Brenda Lee.

has established preserves throughout the country to protect the unique native flora and fauna, and importantly, they have set the goal of protecting more than 30% of the entire country by 2030.

With the goal of finding the ancestral carriers of the deadly hantavirus, in 1999 a group of American and Mongolian scientists established a series of remote field bases to collaborate on a survey of native mammals. Among the many small mammals they trapped and collected, one of the most abundant was the Ural field mouse, *Apodemus uralensis*, whose overflowing populations blanket the Mongolian grasslands. The survey would take multiple expeditions—five in total—to begin to map the distribution of these common rodents.

Field mice are hosts to many kinds of parasites. The mice are tiny islands of parasite diversity, harboring worms, mites, lice, ticks, and fleas, the famous vectors of bubonic plague. In their short life—two years is old age in these mammals—they are eaten by a remarkable diversity of predators. On the Mongolian steppes, corsac foxes, red foxes, gray wolves, and badgers relentlessly hunt the field mice. There are also avian predators, such as black kites, peregrine falcons, and sakar falcons. Throughout the night, 12 local species of owls use their remarkable vision and precise hearing to locate their rodent prey. And during the day, the mice become tasty hors d'oeuvres for nesting cranes, such as the Siberian, white-naped, common, hooded, and red-crowned cranes. Native shrikes stick the mice on branches while they go back for a second course, and ravens pick the mice from the grasslands like peanuts from a giant bowl. With this abundance of predators, the likelihood of a mouse surviving long enough to die from its parasites is very low.

On their first trip in 1999, the scientists from Nebraska and New Mexico were met by their Mongolian colleague Batsaikhan Nyamsuren, zoology professor at the National University of Mongolia. Batsaikhan is known as the "E. O. Wilson of Mongolia," a biologist revered for his knowledge of the natural history of the Mongolian grasslands. Among the Americans were parasitologist Scott L. Gardner, from the University of Nebraska, and mammalogist Terry L. Yates, from the University of New Mexico.

The Americans landed at dawn in Ulaanbaatar, the capital of Mongolia, at that time a city of fewer than a million people. Since it was just after the Soviets had ended their half-century occupation, the city was left without much transportation or infrastructure. The Mongolian scientists and an advance crew of U.S. graduate students had rented a UAZ, a Soviet four-wheel-drive van, and a GAZ 66, a Soviet-made four-wheel-drive weapons carrier. They picked up the scientists at the airport in the van, the interior of which was full of numerous species of biting flies—horse flies, deer flies, and huge moose flies as big as a thumb. The flies, trapped in the van, bounced against the windows trying to get out, but the windows were sealed shut, and the scientists spent the 60 km drive in tortured silence.

The team drove to nearby Gorkhi Terelj National Park, where Batsaikhan and students had set up a field camp. The park, third largest in Mongolia and nearest to the capital, welcomes visitors for hiking. The group was relieved to arrive among fields of blue lupine surrounded by dark green larch intermingled with birch trees. The plan, originally devised by Batsaikhan and his colleague, Dr. Ganzorig Sumiya, parasitologist and professor at Mongolian National University, was to set up a long-term international ecological research site, so the location for the trapping had to be in an undisturbed protected area without easy public access.

The location of the trapping grid was 1,000 feet up a mountain. The scientists carried traps up the steep stairway-like trail, and along the way, munched on khuushuur, a Mongolian empanada made with camel meat, onions, and spices. They picked their way through talus rockslides covered in wildflowers and teeming with pikas screeching at being disturbed. At the top, the scientists set up a trapping web: a gigantic 1,000 meter wheel, with each of eight spokes running to a central hub, with live traps set along the spokes every 10 meters. This pattern of trapping allowed the scientists to estimate both the density and the number of individuals of different species of small mammals in the region. A computer program had been designed to estimate population features using the same pattern of trapping across different long-term ecological research sites. Data obtained from the site in Mongolia would eventually be compared with those from across the globe in New Mexico.

Once the traps were set, the team skidded back down to the campsite at the base of the mountain. Soon a routine was established: climbing to the site each morning, collecting the animals from the traps, and bringing them back to the camp for processing. Surrounded by fields of wildflowers with an occasional yak grunting nearby, the scientists set up a field dissecting laboratory. Gloves, face mask, and eye protection were standard and occasionally full body suits were necessary. Processing was a laborious assembly line: at the time of collection, each mammal received a unique scientific number that stayed with the specimen and its parasites and tissues (like heart and liver) for the rest of its scientific life. The parasites living on the outside of each small mammal—the ectoparasites—were shaken loose in a plastic bag and preserved in ethanol for later study. The search for the internal, or endoparasites, proceeded in the alimentary canal, the body cavity, under the skin, and in every organ, including the eyes, brain, and bladder.

At night, the field camp looked like a zombie movie set—a glowing tent inhabited by featureless apparitions hunched over tables of guts and Petri dishes surrounded by green coils of smoking mosquito repellent. A generator, humming continuously, provided bright lights for the camp site and the laboratory. Every few minutes, a centrifuge spun bone marrow into its cellular components, while the researchers dripped fixative to preserve the material. The scientists then extracted a drop of cells isolated from the bone marrow onto a microscope slide, which when stained, revealed the chromosomes present in the cells. Next the heart, liver, and kidney were extracted from the animal, inserted into cryotubes, and placed, among rising clouds of steam, into a tank of liquid nitrogen.

In the Mongolian field laboratory, the scientists put fecal pellets, packed full of protists, worm eggs, bacteria, and viruses, into vials of potassium dichromate, which preserved the coccidian protists while killing the bacteria. Coccidia were present in every species of mammal that was examined. Coccidia, especially of the genus *Eimeria*, are highly variable—some species are extremely host specific while others are more indiscriminate, inhabiting any gastrointestinal villi they come across. Still others demonstrate host switching, jumping from their normal host to an entirely new animal.

Through the entire process, everyone remained on heightened alert for species never seen before. Like jewelers searching for imperfections in diamonds, the researchers examined the contents of every organ of each mammal. Depending on what was found, a decision was made about the preservation technique. Some parasitic worms, such as tapeworms, trematodes, or acanthocephalans, were placed in distilled water, which relaxes them by creating an osmotic imbalance. Researchers would sort those for morphological study to be preserved in formalin-filled glass vials, while parts of the same worm for molecular study were put into cryotubes in liquid nitrogen or into ethanol-filled vials. The nematodes required treatment in glacial acetic acid, which caused them to uncoil and stretch out, revealing their unusual structures.

Back at the museum, study of these specimens would reveal the genetics of both hosts and microparasites, organisms too small to study in the field, such as viruses, tiny worms, and single-celled protists. Blood and other tissue samples would be carefully transferred to a medical center to undergo what is called ELISA testing. ELISA stands for enzyme-linked immunosorbent assay, a means of detecting the presence of antibodies to viruses, and in this case, those related to hantavirus. Studies of the coccidia and the worms would reveal the value of the parasites as environmental indicators and probes for biodiversity. After their initial visit in 1999, the team regrouped in Mongolia each year between 2009 and 2012.

Sometimes scientists, while researching one question, find answers to others along the way. On their very last visit in 2012, the scientists found a critical clue to another deadly parasite. Camped above Har-Us Lake outside the original silk road city of Khovd in far Western Mongolia, the team set out live-traps, and in one, they found a small mouse-like vole whose liver was packed with larval tapeworms. The scientists immediately recognized the tapeworms as *Echinococcus multilocularis*, whose lobed cysts had replaced more than 90% of the vole's liver. The discovery of this single individual revealed a cascade of ecological information that would serve as the basis for efforts to protect the people of Western Mongolia from a chronic, crippling, and deadly disease. When people are infected with this species of tapeworm and are not treated, about 90% die of the disease.

The tapeworm in the vole confirmed for the first time its presence in Mongolia. The vole is an intermediate host that carries the larval parasites. Transmission occurs when a Mongolian gray wolf is infected with adult tapeworms. The tapeworms produce eggs in the wolf's gut which are eventually passed in the feces. The vole, hungry for calcium, eats some of the mineral-rich feces while accidentally consuming the tapeworm eggs. Inside the vole, the eggs hatch in the gut, and the tiny larvae then move from the gut into the liver. Inside the liver, they begin to form cysts, which sometimes also grow in other organs. When a hungry wolf, fox, or dog eats the vole, the transmission cycle begins again with thousands of new tapeworms growing quickly in the small intestine of the host.

In Mongolia, young kids become infected with the tapeworm from their dogs that eat voles around the family home. Once inside a child, the tapeworm creates silent havoc, growing inside the liver as the young person develops into an adult. The presence of the tapeworm in children is not usually apparent, but in adults the cysts continue to grow until

Figure 8. *Echinococcus multilocularis* lifeline. Illustration by Brenda Lee.

they begin to produce catastrophic symptoms. Liver function may decrease, causing jaundice; cysts that travel to the brain cause seizures, and others in the lungs leave the person gasping for breath. The discovery of the tapeworm in its wild Mongolian host alerted public health officials that the parasite was endemic to the region. The fact that the entire parasite life cycle occurred around the region of the lake meant that health measures, including treatments that destroy the adult parasites in the dogs before they move to children, could become standard practice.

The research in Mongolia raised many questions that the scientists are continuing to study. Although the research team greatly expanded the knowledge of the diversity of parasites in rodents and other mammals in Mongolia, the puzzle of the origins of hantavirus remains largely unsolved. The good news is that hantavirus has not yet been found in Mongolian mice, and there remains hope for containment of the spread of disease. Professors Batsaikhan and Ganzorig Sumiya at National University of Mongolia are world leaders in Mongolian mammalogy, parasitology, and ecology. The Mongolian student, Altangerel Tsogtsaikhan Dursahinhan, studied for his PhD with Scott Gardner at the Manter Lab, and is training to represent the next generation of parasitology curators in Mongolia.

Field parasitologists are modern-day explorers. Like the early natural historians who searched unexplored wilderness for new plants and animals, parasitologists typically find new species on and in animals that are already known. Even the most familiar animals harbor worlds of biodiversity that are only beginning to be understood. As parasitologists discover new species, they map their life cycles through space and time, describing the webs of interaction between a single species and its various hosts. This understanding provides a road map for how humans can understand, and ultimately live with, the diversity of parasites in the world.

Chapter Ten

Parasites in Paradise

Parasites can have catastrophic impacts, particularly on the human populations who live on isolated islands. Located between Russia and Alaska in the Bering Sea, the island of St. Lawrence spans about 160 km, about the size of Isabela, the largest of the Galápagos Islands. Like the Galápagos, St. Lawrence Island is volcanic in origin. Although politically part of Alaska, St. Lawrence Island is geographically closer to the Chukchi Peninsula of eastern Russia, and it is a remnant of the land bridge that once connected Asia to the Americas.

St. Lawrence Island has been occupied for thousands of years by the Siberian Yupik People, who share a common language and culture with the Indigenous tribes of the Chukchi Peninsula. In modern times the island continues to be a harsh homeland: villagers who traditionally made their living from fishing and hunting walrus and whale barely survived the cycles of famine over the years. Around 1900, reindeer were introduced to alleviate periodic food scarcities, and a large herd is now well established on the island. Siberian Yupik daily life used to center on the use of dog sleds for transportation, mail delivery, hunting, and herding. Dog sledding has been a feature of Yupik culture for hundreds of years, and the dogs and their owners have enduring bonds of friendship and respect.

In the late 1940s, veterinary doctor and scientist Robert Lloyd Rausch began a survey of parasites and mammals on St. Lawrence Island for the Arctic Health Research Center, part of the U.S. Public Health Service. Rausch was no ordinary veterinarian: he had studied mammalogy, herpetology, and entomology as an undergraduate and eventually

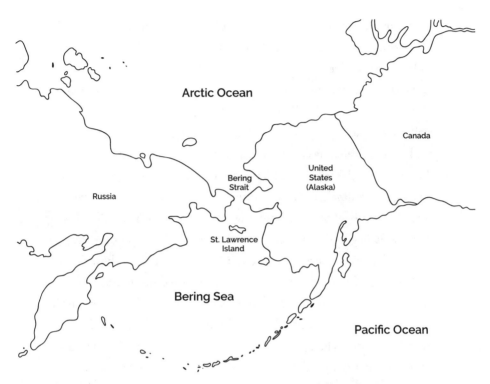

Map 2. Map of St. Lawrence Island and the Bering Strait. Illustration by Brenda Lee.

returned to graduate school to study parasitology and wildlife management. Rausch was a gentle, humble iconoclast who understood the deep connections between the health of people and that of other animals. His research foreshadowed the entire field of modern veterinary medicine and the contemporary concept of One Health, the international and interdisciplinary collaborations that focus on the connections among people, other animals, plants, and the environment, and their role in global health. Rausch arrived on St. Lawrence Island intending to document diseases carried by mammals on the island and to understand their threats to Native populations. Rausch was a keen listener, grateful for the hospitality and assistance provided to him by the Siberian Yupik villagers. And with their help, the quiet biologist and his team uncovered an insidious cycle of transmission that involved a tapeworm, native voles, arctic foxes, the invaluable sled dogs, and the villagers themselves.

Predators and their prey have always undergone reciprocal population cycles, but on islands and especially in the Arctic, these spurts and collapses impact nearly every other species in the region. Sometimes the cycles are triggered by climate and sometimes by increased food sources. Regardless of the trigger, suddenly more individuals of a rapidly reproducing prey animal, like a vole, survive, and they seem to be everywhere, and the numbers of their predators, with the bursting availability of food, soon begin to increase. When this happens, it is not only the prey and predators that become abundant; Rausch was able to show that the parasites also follow this pattern. When he started his surveys, the populations of the tundra voles were reasonably stable. Sometime in the early 1950s, the voles began to wildly increase in numbers, far more than the residents had previously observed, and soon so did their predators, the arctic fox. By the winter of 1955, all foxes that were tested were found infected with the tapeworm *Echinococcus multilocularis*. Infection did not mean just a few tapeworms; a light infection might mean 25,000 individuals in a single fox, and a heavily infected animal might have as many of 450,000.

Echinococcus multilocularis is a tiny worm that lives out its life in two kinds of hosts: an intermediate rodent host, like a vole, and in a definitive carnivore host, like an arctic fox. Rausch suspected that the tapeworm may have been brought to St. Lawrence Island from Asia by foxes traveling on the pack ice. Before the arrival of humans, the tapeworms might have quietly lived out their life cycle within the rhythms of predator-prey cycles or, in this case, vole-fox population fluctuations. But Native people brought dogs to St. Lawrence Island, and since dogs are closely related to foxes, they too began to serve as definitive hosts. Although humans are not a natural host for the tapeworm, the islanders, living in close contact with their dogs soon became infected.

Infected villagers began to show alveolar hydatid disease, the signs of a chronic infection with the tapeworm. Although humans are an accidental host, the effects of infection are pathological for both the person and the larval tapeworm. When the tapeworm eggs are ingested by a person, they start the larval stage as they would in a vole, and the infection might go unnoticed for years as the cysts slowly enlarge, disrupting normal functioning of the liver, lungs, and other tissues. On St. Lawrence Island, at the time of Rausch's studies, more than two-thirds of the human cases were fatal. Today, the Siberian Yupik villagers have many fewer dogs and they regularly deworm the ones they have, so human infection is a rare occurrence.

One kind of parasite or another infects nearly every human being. Over the course of the past century, the science of parasitology has made remarkable progress identifying transmission patterns, creating treatments for infections, and gaining an understanding of what makes some parasites more deadly than others. Finding each missing letter in the parasite crossword puzzle is a step to completing the overall picture of how parasites manage their success. Scientists still don't understand how many parasites choose their hosts and survive under changing environmental conditions. Parasites balance their conservative need to remain with known hosts with an opportunistic streak that enables them to make radical changes under the right conditions. For each parasite species, knowledge of how and when that balance gets tipped adds essential information to figuring out how humans can survive with parasites.

Compared with the nearby mainland, islands nearly always have fewer species and less complex communities of organisms. This was brought to the attention of scientists by the ecologist Robert H. MacArthur, who modeled how the number of species on an island is a function of its size, distance from the mainland, and rates of immigration and extinction. It turns out that islands, particularly those very isolated from mainland continents, are excellent laboratories for studying how parasites make their living. Compared with other archipelagos, the Hawaiian Islands are remote from any mainland continent, and their native fauna evolved in almost complete isolation. About 1,500 years ago, the Hawaiian Islands were discovered by adventurous Polynesian sailors venturing out from other Pacific Islands. Until the settlers arrived, there were only two mammals native to the islands: the Hawaiian monk seal, and a tiny bat known as the Hawaiian hoary bat. Today, both these mammals are in danger of imminent extinction.

Bats have a great diversity of parasites and a unique physiological resistance to very pathogenic ones. This may be an indirect result of the large number of bat species, their ability to fly, or their great evolutionary age. The 1,400 known species of bats make this group, after rodents, the most diverse and widely distributed of all mammals. Bats are the only mammals that have true flight. What are called flying foxes are actually a type of fruit bat, but flying squirrels, unlike bats, only glide from tree to tree and do not actively propel themselves through the air. Viruses have a great affinity for bats, and these flying mammals are known to harbor some of the most dangerous: rabies virus, hantaviruses, Ebola and Marburg viruses, and the SARS-CoV group of viruses. Although bats carry these deadly viruses, their innate and acquired immune systems are thought to protect them from succumbing to disease. Bats have been implicated in playing a role in the transmission of modern pandemics, including the Ebola outbreaks in Africa during 2014 to 2016, and the SARS-CoV-2 global pandemic that emerged in late 2019. Understanding the dynamics between bats and their parasites is at the forefront of research on global health and biological diversity.

The presence of a tapeworm in the rare and isolated Hawaiian hoary bat has raised new challenges for understanding parasite life histories. In the 1970s, Robert Rausch, fresh from his work on St. Lawrence Island and in the Arctic, published a paper describing a tapeworm called *Hymenolepis lasionycteridis* found in the Hawaiian hoary bat. He suspected that Hawaii might originally have been occupied by a single group of migrating bats carried to the islands by winds from North America, although he considered repeated colonization also possible.

The bats would have carried the tapeworms to the islands with them, and over time, the bats and their worms evolved in isolation under changed environmental pressures and became distinct from the mainland species. Not only did the bat become an endemic species, found nowhere else, but the tapeworm did as well.

Scientists modify their predictive explanations when new data no longer fit a paradigm. The survival of the tapeworm in the Hawaiian hoary bat may challenge contemporary theories of parasite transmission. Since it is rare for a tapeworm to lose an intermediate host entirely, the tapeworms that arrived on Hawaii would have had to find a local intermediate host in order to survive. This likely involved a novel and unexpected host switching event, although how this occurred is still a perplexing mystery. Very little is known about the life history of most bat parasites and their viruses, and the story of tapeworms in the Hawaiian hoary bat is full of guesswork and speculation. Studying tapeworms on island faunas might be key to understanding ecological fitting and the extent to which there are limits to the ways in which organisms adapt to change. The complex systems of ecological interactions on islands offer scientists a simplified laboratory for mapping the intersection of evolution, community structure, and ecological diversity. Resolving the life cycle of tapeworms in bats may be one important step to understanding how other organisms, like viruses, jump from one host to another.

Another island parasite puzzle focused on the rice rats of the Galápagos Islands. Collected by Charles Darwin on San Cristóbal Island in 1835, this furry mouselike creature was the first known example of an endemic rodent on the islands. The Galápagos are just far enough away—600 miles—from the coast of Ecuador to have served as a generator that nursed many new species. Sometime about 3 to 4 million years ago, relatives of those rodents survived the journey from the South American mainland, perhaps on a floating log, and landed in the Galápagos. These first individuals reproduced and eventually diversified, giving rise to new species of rice rats.

Rice rats are small, pointy-nosed, generalist foragers that feed on insects, seeds, and other plant parts. On the islands, they help disperse the seeds of native trees, like palo santo. They are hunted by Galápagos hawks and short-eared owls, and their nestlings are known to be killed by the poisonous fangs of the giant Galápagos centipedes. The protective isolation of the Galápagos ended in the seventeenth century when whalers began stopping at the islands to replenish their supplies. And every time a ship docked, it left behind tough thugs from the mainland—rodents like black rats, Norway rats, and house mice. The earliest ships dropped off goats for future food supplies, and these began decimating the island flora. Eventually cats were brought that indiscriminately hunted the naive native rice rats. Today these pressures have begun to drive the native rice rats to extinction.

Competition has always been a mainstay of evolution—it creates some of the selection pressures that push species to become different from one another. But the balance of evolutionary tug-of-wars is skewed on islands. Mainland species are tough gang members, used to defending their territory under difficult conditions. Island species tend to be naive and fragile. Never having had to fight for survival with close competitors, they never evolved the means to defend their lifestyle. When mainland species arrive on islands that had previously been isolated, they have an unfair advantage, and that drives many unique island species to extinction.

In the 1990s, the mammalogist Robert Dowler arrived on the Galápagos to search for evidence of rice rats. At the time, only two of the species that had once inhabited the islands were known. Through tenacious persistence, he found evidence that two more rice rat species, previously thought to be extinct, were still alive. His diligent surveys doubled the number of rice rat species known to have survived on the Galápagos, but all four are still barely hanging on. Dowler was thorough in his research, so he considered not just the rice rats, but also the community of organisms that lived inside them. To learn more about them, he sent specimens to the Manter Laboratory of Parasitology, where they would be studied and eventually identified. But when Curator Scott Gardner examined a specimen of tapeworm from one of Dowler's rice rats, he was baffled. The specimen was clearly a *Railliet-ina* tapeworm, belonging to a genus that has the most species of any group of tapeworms—at least 750 are known globally. But how did the rice rat tapeworm get to the Galápagos Islands?

Gardner wrestled with how to make the task of identification more manageable. The samples were in rough shape with no usable DNA, so genetic barcode identification techniques could not be used. It would have to come down to the laborious task of comparative morphology, examining how the detailed features of this particular tapeworm matched up to the features of hundreds of others. Gardner considered the most likely scenarios: did the tapeworm come millions of years ago with the original rice rats that arrived in the Galápagos? Or was this a more recent introduction, transmitted from rats and mice that arrived with ships? Could this be a case where the parasite switched hosts from something like a bird to rice rats?

It is entirely possible that when the rice rats originally arrived on the Galápagos, they brought with them freeloaders in the form of ancestral *Raillietina* tapeworms. This scenario would point to present-day South American *Raillietina* as likely close relatives to the Galápagos species. Of the known South American species, one kind is found only in monkeys. Another is located mainly on the Caribbean side of the continent, blocked by mountains from an easy route to the island. Gardner started his search with records of *Raillietina* in Ecuador, the closest country to the Galápagos, and he focused on those that might infect birds or rodents, because no monkeys are known to have made the trip.

There were other hypotheses that had to be considered as well. Most of the ships to arrive at the Galápagos came directly from either South America or Asia. Boats from Southeast Asia routinely stopped at the

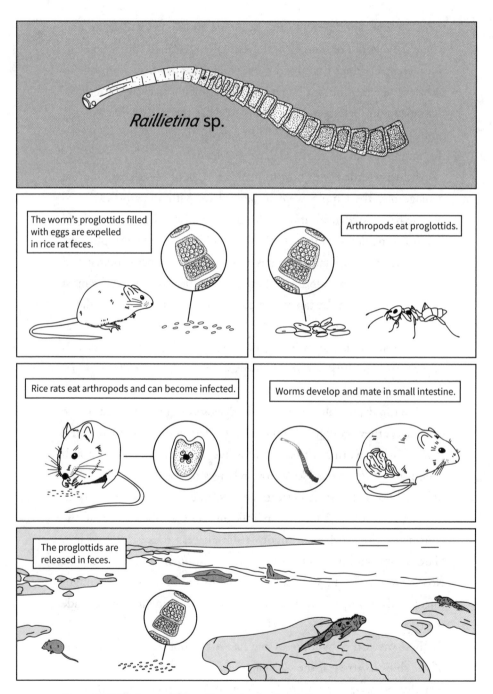

Figure 9. *Raillietina* sp. lifeline. Illustration by Brenda Lee.

islands in the 1700s to harvest tortoise meat, and in the process, they dropped off rats that then bullied their way into the food web. If these rats harbored *Raillietina* tapeworms, then the worm could have switched hosts and settled in the local rice rats. This meant that the endemic tapeworm could be most closely related to the *Raillietina* species in Asia.

A third hypothesis was based on the observation that at least one species of *Raillietina* uses ants as an intermediate host. There are 12 species of ants endemic to the Galápagos, and the islands' famous finches may have served as a definitive host for the tapeworm. It is conceivable that at some point in history, the tapeworm engaged in what is now referred to as *ecological fitting*. Under just the right environmental pressures, the tapeworm modified its infection pathway to go from ants to rodents, instead of from ants to birds. The origin of the *Raillietina* tapeworms in the Galápagos rice rats remains a mystery yet to be solved by parasitologists as they collect additional molecular information to assess these competing hypotheses.

Parasitologists have long traced the tightly correlated systems of parasites and hosts. Often parasites evolve in concordance with their hosts. If a definitive host moves to a new location, its parasites must find a suitable intermediate host in the new environment in order to survive. Sometimes the parasite gets lucky and makes do with the next best thing, an intermediate host closely related to the original. On St. Lawrence Island, when the arctic foxes brought with them tapeworms, the parasites easily established themselves in the local voles, the same species as those to which they had been accustomed. And during a population explosion of voles, the tapeworms also made their way into a new, but related definitive host, the sled dogs. On the Galápagos, the tapeworms in the rice rats probably switched intermediate hosts from one species of ant to another, and they may have also switched their definitive host from black rats to the native rice rats. These represent small manageable jumps from one species to another, all in the process of adapting to a changed environment. Such gradual stepwise ecological fitting is a way of life for parasites in an unstable world.

Chapter Eleven

Diversity in the Dunes

Parasites are central players in the ecology of communities. In any ecosystem, one can never really understand the web of feeding relationships unless parasites are included. Parasites are consumers that typically infect organisms larger than themselves. Sometimes multiple parasites inhabit the same hosts, and they can form their own kind of community. The scientists at the Manter Laboratory of Parasitology have been attempting to fill in the gaps in the web of community interactions among the organisms that live in the Nebraska Sandhills.

Stretching across north-central Nebraska and encompassing a quarter of the state, the Nebraska Sandhills is one of Earth's largest regions of mixed prairie on stabilized sand dunes. Ancestral rivers running from the Rocky Mountains originally deposited the sand, which then was shaped by winds and stabilized by grasses. Between the rolling hills of dunes and wet marshes, 1,000 small shallow lakes dot the region. More than 300 different species of birds, mammals, reptiles, and fishes make their living in this unique ecosystem. Coyotes, weasels, and red foxes hunt, scavenge, and harass the numerous voles and mice in the area. Mule and white-tailed deer and pronghorn antelope forage relentlessly on the more than 700 species of grasses and forbs. Black-footed ferrets, badgers, raccoons, opossums, and skunks opportunistically pluck small animals and insects. Prairie chickens, horned larks, and sandhill cranes scrounge for seeds and insects. Hognose snakes feed on toads and frogs, while rattlesnakes consume small mammals. Small fish such as killifish, bullhead, and perch struggle to survive in the alkaline lakes. In the evening twilight, seven species of owls and eleven species of bats claim

Map 3. Map of Nebraska Sandhills. Illustration by Brenda Lee.

ownership of the skies. Among this extraordinary diversity, parasites have staked their claim alongside nearly every node in the web of interactions.

Parasites in the Sandhills occur in all species of the rodents, the largest and most diverse group of resident mammals. Aboveground, beavers and muskrats carve up streams, prairie dogs socialize, kangaroo rats hop from snakes, and voles and deer mice make runways in the grass. Rodents form the foundation of the carnivore food chain on land. They are bucktoothed champions: all are characterized by large continuously growing incisors—two upper and two lower—and a large space between them and the cheek teeth, since they have no canines. They symbolize a resilient lifestyle, descendants of the mammals that survived the cataclysmic impact at the end of the Cretaceous. Small size and the ability to reproduce three to four times a year is the way to be tough enough to survive in hard times. Their hardiness is reflected in the fact that about 40% of all living mammals are rodents.

Among the rodents living in the Sandhills, grasshopper mice stand out. They munch on a wide variety of insects, including beetles and crickets. Grasshopper mice readily ambush other species of mice, grabbing them at the back of the head and, using their incisors as stilettos, pierce their victims' skulls. The diverse predatory methods used by grasshopper mice are apparently learned early in life. They have a tough stomach lining, an adaptation to eating indigestible arthropod chitin. Grasshopper mice are distributed widely across windblown sand dunes among prairie grasses and dry-land shrubs, where they communicate with each other with odd vocalizations, including high-pitched chirps and sharp whistles. In the winter months when insects are scarce, grasshopper mice don't hesitate to feed on grasses and seeds. They are open-minded about their dens, taking over an abandoned gopher burrow when available and willing to dig their own when the occasion demands. Once situated, their homes have an elaborate floor plan: a nest burrow that can be sealed against moisture, a cache burrow for seed storage, a defecation burrow for waste, and even short sign-post burrows, tinged with warning pheromones that announce territorial boundaries.

When scientists surveyed the Sandhills for parasites, they were surprised to find a previously unknown species of tapeworm living inside this savage little predator. The specimen was deposited in the Manter Laboratory of Parasitology under the name *Hymenolepis diminuta*, a common tapeworm found in rodents throughout the world. One doesn't always know what's special about a species until later, when it gets re-examined. When the same tapeworm was seen in other grasshopper mice, they began to suspect it was something new. The new species, *Hymenolepis robertrauschi*, was described by Gardner and named for the notable parasitologist Robert Rausch. *Hymenolepis robertrauschi* is now known to range from Nebraska through the Great Plains into New Mexico, wherever its host, the northern grasshopper mouse, is found.

How does a tapeworm generate a mystery story? The answer is, when clues to its lifestyle keep scientists guessing—in this case, the search for the intermediate host of *H. robertrauschi*. The eggs of the common tapeworm, *H. diminuta*, develop in mealworms, the larvae of the darkling beetle *Tenebrio molitor*. The mealworms serve as the intermediate host, passing on the larvae when they are consumed by a rodent, the definitive host. But *H. robertrauschi* eggs don't develop well in mealworms, so there must be another intermediate host that is more suitable. That host has yet to be identified.

The mysteries don't end there. Northern grasshopper mice often live in abandoned pocket gopher burrows, but they carry an entirely different community of parasites. Grasshopper mice harbor tapeworms, nematode worms in their intestine, and single-celled protists called *Eimeria* in the lining of their large intestines. Even though the pocket gophers in the Sandhills live with and around grasshopper mice, they share no species of parasites. To solve what is keeping these two mammal hosts from sharing their parasites, parasitologists examined the lifestyles of the host species and the community of organisms they interact with. The

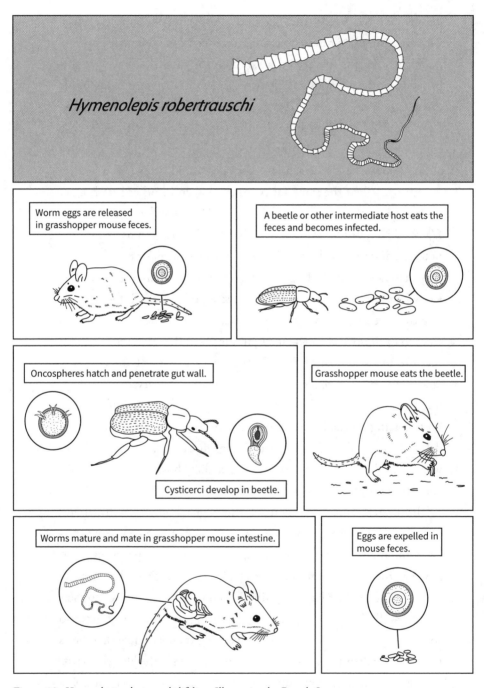

Figure 10. *Hymenolepis robertrauschi* lifeline. Illustration by Brenda Lee.

central player in this community turned out to be pocket gophers, and to understand pocket gophers is to go underground.

Aboveground, the Sandhills may seem like a busy place, but that pales in comparison with what goes on below the surface. Beneath the once-churning sands, some 50 species of small mammals make their homes, and like subway construction workers, they continuously search, dig, and expand their burrows through the soil. The champion diggers of all the mammals in the Sandhills are the pocket gophers, tough little mammals that have the ability to turn into hyperactive subterranean plows. Pocket gophers, by one estimate, turn over every inch of the Sandhills 10 to 20 times over the course of a few thousand years. The region may be the largest unplowed prairie in the world, but below and on the surface, soil is continuously being churned.

There are 41 species of pocket gophers known. They range from Canada south to the banks of the Rio Atrato in Colombia. These are part of an ancestral group of rodents that colonized North America and diversified into aboveground species, such as kangaroo rats and pocket mice, and the subterranean pocket gophers. All pocket gophers were originally native to North America, and they all have characteristic fur-lined cheek pouches that are used to transport food.

An observer stationed above the sandhills will see random pop-up mounds of soil—first in one place, then in another. These are the subtle opportunistic entrances and exits that link together vast networks of burrows. After a while, the observer might also see a grass stem begin to wiggle and then suddenly disappear, pulled down by a foraging gopher. As their moving colonies plow up soil over thousands of square miles, pocket gophers provide deep fertilization and aeration for the Sandhills' soil. In this way, the gophers aid the slow succession of prairie plants that adapt to ever-changing environmental conditions.

Pocket gophers construct their subterranean burrows parallel to the soil surface at depths ranging from a few centimeters to more than half a meter. Most of their lives are spent underground, since the burrows provide safe places to raise their young and foraging sites where roots and plants can be grabbed from below. A typical system has a single long continuous tunnel with smaller sections that branch out. There are shallow tunnels located below roots of grasses and other forage plants, and deeper tunnels with stable temperatures and moisture for nest cavities and food caches. There are tunnels with latrines, drainage outlets, and ventilation chimneys. Burrows form a three-dimensional network in which tunnels are separated from one another by only a few centimeters. The burrows are dynamic structures, with new excavations and plugging of old burrows going on day and night. These movements reflect the underground lives of gophers: establishing new food sites, defending territories, and raising young. During winter, even when the ground is frozen, the gophers stay active below. Their most vulnerable time is when the young and adults disperse, moving out on moonless nights to search for mates and establish new burrows. Owl pellets contain large numbers of gophers of all ages, attesting to the extraordinary risk of leaving the natal burrow systems.

Pocket gophers are zookeepers that carry a literal tree-of-life of parasites. They harbor fleas, lice, and mites on their skin and fur, single-celled *Eimeria* protists in their intestinal lining—and in their gut, multiple species of tapeworms, nematodes, and even thorny-headed worms. Most of these freeloaders don't cause lasting harm to their furry host. Too many mites are an exception: they make the pocket gophers itch, and their incessant scratching can lead to mange. Like many mammals, they are inhabited by so many other organisms that the concept of an individual seems like a misconception. There are more viruses and bacteria living on and in a single pocket gopher, than there are pocket gophers in all of the Sandhills. And this isn't just true of the microbes—take any one group of parasites found in pocket gophers, nematodes for example, and it will lead you to an entire family of related species, each of which has carved out its own niche inside the host.

If a single Sandhills pocket gopher represented a house, then one species of nematode would be found in the living room, another in the kitchen, and still others in each bedroom. One species might wiggle through the dog door, while another might wait in the garden for someone to pick it up. In fact, each species of nematode has its preferred location in a pocket gopher. One kind is at home in the stomach, another prefers the small intestine, still others like the cecum, the small pouch between the small and large intestine, and other species inhabit the body cavity. With so many nematodes throughout its body, one might suspect this would be a very sick pocket gopher. But most parasitic nematodes live quietly inside other organisms.

Pocket gophers and parasites have, over many generations, gotten used to one another, and in some cases, depend on each other for a healthy life. For example, the nematode *Ransomus rodentorum* occurs in the cecum of almost all pocket gophers and helps them use anaerobic digestion to break down plant foods the gopher ate. The filarioid nematodes that live in the body cavity of pocket gophers have a symbiont or partner, a bacteria called *Wolbachia*, to help them absorb nutrients. Most filarioid nematodes have evolved a dependence on *Wolbachia* and cannot reproduce and survive without them. Each species of nematode has its own species of *Wolbachia*. Once comfortably inside the pocket gopher, a nematode and its *Wolbachia* are like a pair of unassuming guests who've earned the invitation to stay around.

Although grasshopper mice utilize the deserted burrows of pocket gophers, it remains a puzzle why there are no parasites shared between these two rodents. Apparently, pocket gophers and grasshopper mice are like next door neighbors with different job schedules; gophers and grasshopper mice rarely come into contact. In every sense of the term, the two species have distinct niches, sharing habitat but utilizing it in entirely different ways. Grasshopper mice spend more time on the surface hunting down insects, while pocket gophers focus on pulling plants from their underground refuges. These differences in behavior and ecology are enough to shape their relationships with parasites, forming entirely distinct collections of species. In this way, the historical host specificity of the parasites, instead of gradually changing over time, is robustly maintained by the differences in ecology. If we consider the niches occupied by the parasites, rather than by their mammalian hosts, they begin to look even more different. For a parasite that lives in a grasshopper mouse gut, the regular influx of dead insects provides an entirely different niche from the plants in the gopher's gut. In this way, parasites and their hosts form a mini-ecological system of interwoven niches in time and space.

The Nebraska Sandhills is a landscape in perpetual flux. Shaped by past geological events, the dunes and grasslands are continuously formed and reformed by rain, drought, and wind. The hosts of parasites stay sensitive to these changes, shifting their local food preferences, moving to new ranges, or going extinct. The parasites of the mammals track these changes, adjusting where they can, moving with their hosts to new locations or switching to new hosts. Today the short-term changes in weather are embedded in a much longer-term process of climate change. The yearly cycles of wet, dry, and windy, familiar to the Sandhills rodents and their parasites, are becoming less predictable. This presents a new set of forces to be reckoned with, whether as anonymous cohabitants of the same tunnel or as the parasites and their hosts, bound together by their ecology and their evolutionary past.

Chapter Twelve

Kissing Bugs and Bucktoothed Potatoes

Parasites are often particular about their hosts, which makes them classic examples of ecological specialists. And many species are host specific, never veering from their initial comfort zone. But parasites possess more than one strategy for survival. Parasitism is a lifestyle that has evolved in diverse organisms that, themselves, are constantly evolving. While some parasites are extreme specialists, others are more flexible in their habits. Parasites are known to survive the extinction of their hosts by opportunistically switching at just the right time. Sometimes a parasite is found in a new location because it was carried along as its host moved to a new region. Other times, parasites gradually move from one host to another, expanding their reach when the opportunity presents itself. One thing is certain: parasites are rarely static entities that stay with the same host and in the same region throughout their evolutionary history.

One famous example from South America is the parasitic flagellate *Trypanosoma cruzi*, a single-celled protist that moves with a whiplike flagellum and lives in blood and other organs of mammals, including humans. A common parasite in many species of mammals throughout South and Central America, *T. cruzi* is spread primarily by kissing bugs, members of the Reduviidae, a family of blood-sucking insects. The parasite is transmitted in many ways—directly through the bug's feces, on contaminated fruits, through blood transfusions or organ transplants, or from mother to fetus. Infection results in Chagas disease, a debilitating ailment that affects 8 million people in Latin America and each year leads to thousands of deaths.

Chagas can persist unrecognized for years. Typically, a person is bitten by a kissing bug that takes a full blood meal and while it is feeding it defecates on the host. The infectious forms of the parasite are then transmitted to the host via scratching or through mucus membranes. An infected person retains the parasite for a very long time, and the infection rarely clears without treatment, and even then, the heart and other organs may be severely damaged. A famous sufferer was Charles Darwin, believed to have been infected during his South American explorations, who then lived with the effects of the disease until he died. There is not yet a vaccine against infection; effective treatments work only in the early stages and not when the infection is chronic. Although Chagas is pervasive in many South American countries, much is being done to eradicate the disease by conducting massive programs for vector control, diagnosis, and treatment.

The evolution of the Chagas parasite goes back to an ancestor from Gondwana, the ancient southern supercontinent that existed around 300 million years ago. As South America formed, the relatives of the *T. cruzi* parasite probably infected fish and amphibians, and later diversified into forms that infect mammals such as opossums, armadillos, squirrels, bats, and monkeys. When ancient humans arrived in South America more than 10,000 years ago, their living conditions in caves and primitive dwellings made from mud and plant materials provided ideal hiding and living spaces for kissing bugs. Evidence that *T. cruzi* infected humans early on in South America comes from its presence in a human mummy from the Chinchorro, a 9,000-year-old archaeological site in Chile.

Since the reduviid bugs prefer a warm and humid climate, global warming is enabling them to expand their range into new areas. In the past several decades, the range for the reduviid kissing bugs has been moving north, and the presence of *T. cruzi* is now established in the United States. Antibodies to *T. cruzi* have been detected in the United States in woodrats, and to a lesser extent in striped skunks, raccoons, cotton rats, rock squirrels, black rats, and various mice, including house mice and deer mice. In 2014 there were 11 species of kissing bugs in the United States. Draw a horizontal line through the center of the United States, and all the states south of the line have kissing bugs, and most of the bugs harbor the *T. cruzi* parasite. This means the bugs and their parasite are found in Utah, but not Idaho; Pennsylvania, but not New York; Kansas, and only recently in Nebraska.

At first glance, this seems to be a somewhat simple case of the parasite being carried along with its insect vector as it expands its range northward to new territories. But it turns out that *T. cruzi* has probably been quietly residing in the southern United States for a relatively long time. A 3,000-year-old human mummy from the Rio Grande region of Texas was found to have been infected with *T. cruzi*, most likely being infected by the kissing bugs living in close proximity to the people inhabiting caves in the area along with their normal mammalian hosts, the woodrats. Texas was probably a marginal environment for the parasite, where it could eke out an existence, but not thrive and expand. But climate change is gradually chipping away at the limitations on the parasite's expansion. There is every reason to suppose that Chagas disease cases will increase in the United States as the climate gets warmer and kissing bugs find more suitable habitats and as they move further north.

Clues to the geographic distribution of parasites often lie in the complexities of the movements of their hosts. Rarely do potential hosts originate in one location and then stay put throughout geologic time. Continents are in perpetual motion, and plants and animals tend to flourish in one place, then find ways to move on when conditions become more or less favorable. Sometimes an accidental transplant leads over time to an entirely new lineage, or a change in climate brings organisms together for the first time. Each of the changes confronted by the potential host species brings a cascade of changes for their parasites.

South America is a most fascinating continent for the study of animal movements, particularly mammalian hosts and their parasites. Its geographic features—the Andes mountains, Patagonian grasslands, Amazon rainforest, and valleys and ridges bisected by rivers—isolate populations and generate new species, forming hot spots of biodiversity. About 100 million years ago, South America broke away from the southern supercontinent Gondwana, moving west to become an island, and remained that way for 40 million years. Mammals flourished in island South America, forming a unique and diverse fauna dominated by marsupials, distinguished by their teeth and ability to raise their relatively undeveloped young in pouches. There are more than 100 species of marsupials living today in South and Central America.

Alongside the marsupials were a heterogenous collection of placental mammals, such as the xenarthrans—anteaters, tree sloths, and armadillos. These early placental mammals included the glyptodont, possibly the most heavily armored mammal that ever lived, and *Megatherium*, the giant ground sloth that, when standing, was more than 3 meters tall and could walk bipedally like humans. The remarkable biological diversity of South American mammals expanded as the continents continued to move and global sea levels rose and fell. About 38 million years ago, the distance between South America and Africa was only about 1,500 km, just a little more than the distance from the Galápagos Islands to mainland South America. And as with the fauna that populated the Galápagos, a variety of opportunistic creatures landed in the New World, where they survived and reproduced, eventually diversifying into an astounding array of new mammal species.

Two very different kinds of mammals arrived in South America from Africa by this route. A group of primates with long tails and flat noses made their homes among the canopies of the abundant forests and, over time, gave rise to an entire lineage of South American monkeys. Where the primate immigrants inhabited trees, another group of recently arrived mammals stayed on the ground. These rodents, called caviomorphs, speciated into every niche in South America. Nutria and capybara are at home wading in shallow waters, pacas and agoutis scutter on forest floors, and strange little tuco-tucos make their living digging in burrows. For the most part, the African waifs fit into ecological gaps in the fauna, greatly increasing the diversity of South American mammals.

The newly arrived primates and rodents enriched the South America fauna, but an even more massive ecological blending was still to come. Sometime before 3 million years ago, continental movement formed a land bridge between North and South America, now called the Isthmus of Panama. This land connection afforded a gateway for faunal exchanges between the Americas: opossums, porcupines, armadillos, and jaguars drifted north, while deer, bears, dogs, llamas, squirrels, elephants, and mice came south. In the resulting ecological scramble, the caviomorphs did just fine, and they continued to diversify into numerous species.

The history of South American mammals represents one of the best-known examples of the evolution of an entire fauna in relative isolation and the subsequent upheavals as geological changes enable the introduction of new species. The distribution of parasites, however, depends not only on their own adaptations, but also on the availability of suitable hosts. As the South American mammals diversified, their parasites expanded into new host niches. While the story of the South American mammals is now beginning to be understood, the corresponding narrative of their parasites has been virtually unexplored.

In 1984, Scott Gardner, a 28-year-old graduate student, joined the field crew for a National Science Foundation–funded survey of the mammals of Bolivia. The team was led by two renowned scientists: Sydney Anderson, Curator of Mammals of the American Museum of Natural History in New York City, and Terry L. Yates, Curator of Mammals at the Museum of Southwestern Biology at the University of New Mexico. A decade later, Yates would become famous for discovering the

Map 4. Map of Bolivia. Illustration by Brenda Lee.

source of the deadly hantavirus in the southwestern United States. Not much was known at the time about the mammals of Bolivia, and the team initially collaborated with Gaston Bejarano and Armando Cardozo, scientists at the Bolivian Academy of Sciences in La Paz. The expedition's first field camp was established at Rancho Huancaroma, a large dairy farm two hours southwest of La Paz. The scientists set up their laboratory in a local school on the farm and shared their science with the regular influx of curious kids. Working together over a 20-year period, the team would help to quadruple the number of known mammals in Bolivia to the 410 species known today.

Most of the team spent their time trapping and preparing mammals, but the young graduate student had an additional interest, focusing on the parasites that were showing up inside the mammals that were collected. Gardner chose to zero in on the parasites that lived inside small caviomorph rodents called tuco-tucos that looked like little bucktoothed potatoes. Like pocket gophers in North America, these small diggers excavate their huge burrow systems, creating subterranean passageways where they live nearly their entire lives, feeding on roots and shoots. They have a peculiar mode of digging: they use their clawed front feet for scratch digging, while their front teeth chisel away at hard soil. Their back feet have hair modified into stiff bristles that help them push loose soil around and out of their burrows, like a backward bulldozer. Tuco-tucos are sometimes compared with the Darwin finches in the Galápagos, since each of the 70 described species of tuco-tuco shows environmental preferences and physical modifications—some excavate in lowland grasslands, others along high mountain slopes. And genetically, they are extremely diverse—one species has as few as 10 chromosomes, while others have as many as 70. But one thing that all tuco-tucos share is the abundance of parasites that live with them.

When Gardner opened the first tuco-tuco, he noticed worms inside the animal's cecum, the expanded portion of its large intestine where it decomposes cellulose from roots and plants. He thought at first the intruders were pinworms, a common kind of nematode, because they had the characteristic internal structure of a turkey-baster—a large bulb-like esophagus. In fact, Gardner had come across one of the rarest nematodes, known at the time by only a single species, *Paraspidodera uncinata*. Gardner went on to write his PhD dissertation about the *Paraspidodera* nematodes. Later he surveyed all known species of

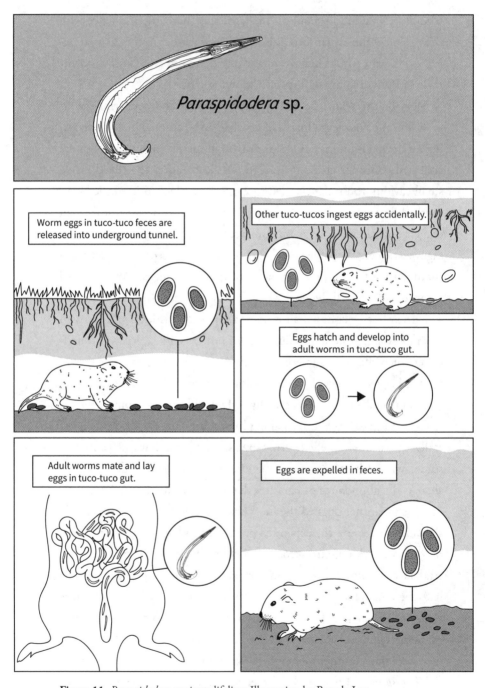

Figure 11. *Paraspidodera uncinata* lifeline. Illustration by Brenda Lee.

Plate 1. Hookworm dispensaries, like this one from 1905, educated the public and treated hookworm infection. Rockefeller Archive. Special Collections, USDA National Agricultural Library.

Plate 2. The broad fish tapeworm, *Diphyllobothrium* sp., that infected this Alaskan brown bear, escapes out of the host's anus. 2007. Photograph by Scott Davis. Used with permission.

Plate 3. This 35-foot-long broad fish tapeworm, *Diphyllobothrium* sp., was retrieved by Scott Davis from the feces of an Alaskan brown bear. 2007. Photograph by Kathy Davis. Used with permission.

Plate 4. Eric Hoberg, chief curator of the former U.S. National Parasite Collection, examines a jar of preserved nematodes from raccoons. 2010. Photograph by Peggy Greb. Used with permission.

Plate 5. This *Stagnicola elodes* snail from a cattle tank on Arapaho Prairie in the Nebraska Sandhills is the intermediate host of trematodes that infect water birds, such as mallards and blue-winged teals. Photograph by Scott L. Gardner. Used with permission.

Plate 6. These *Schistosoma mansoni* flukes from a laboratory mouse show the male and female worms in copula, where they will remain for the lifetime of the host. 2018. Photograph by Gabor Racz. Used with permission.

Plate 7. The *Biomphalaria* snail, shown here, is the intermediate host for the fluke *Schistosoma mansoni*. 2015. Photograph by C. M. Adema and T. Kennedy, University of New Mexico. Used with permission.

Plate 8. This weaponized acanthocephalan, *Pseudocorysoma constrictum*, in the amphipod *Hyalella azteca* was collected near Crescent Lake National Wildlife Refuge, Nebraska Sandhills. The larval acanthocephalan turns bright orange inside the amphipod host, potentially making it more visible to predators. 2018. Photograph by Scott L. Gardner. Used with permission.

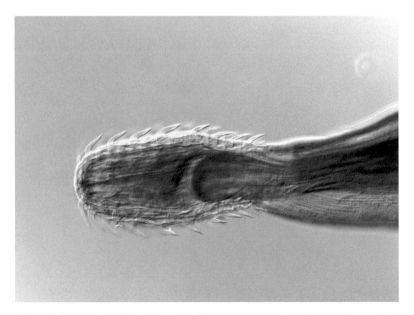

Plate 9. This acanthocephalan, *Polymorphus minutus*, was taken from a mallard duck, its definitive host. Amphipods of the genus *Gammarus* are the intermediate hosts. 2021. Photograph by Gabor Racz. Used with permission.

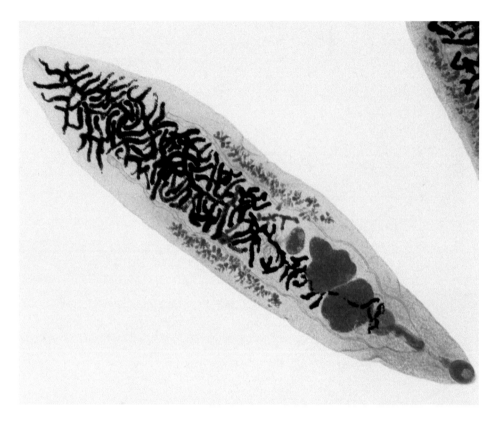

Plate 10. This adult fluke, *Dicrocoelium dendriticum*, lives in the bile ducts of deer or sheep and uses snails as first intermediate hosts. The larvae from the snails are then consumed by carpenter ants, and the parasite induces the ants to climb up grass stems, where they are more likely to be eaten by grazing animals. 2018. Photograph by Scott L. Gardner. Used with permission.

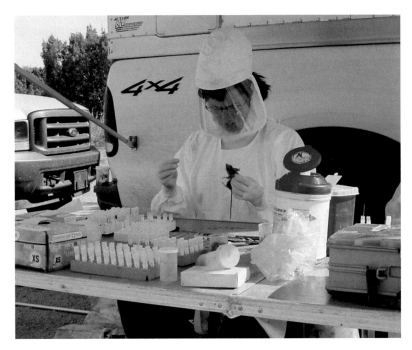

Plate 11. Hantavirus testing of a *Peromyscus* deer mouse in central New Mexico was part of a long-term research program on viral prevalence in small mammals. 2006. Photograph by Gabor Racz. Used with permission.

Plate 12. Bactrian camels are the largest grazing mammals native to Mongolia and the few wild individuals (*Camelus ferus*) that remain are endangered. These, from Khongoryn Els in Gobi Gurvansaikhan National Park, are domesticated Bactrian camels (*Camelus bactrianus*) that descended from a long extinct species that was also native to the region. 2009. Photograph by Gabor Racz. Used with permission.

Plate 13. This field camp near Har Us Nuur was the site of the first confirmed locality of *Echinococcus multilocularis* in an intermediate host in Mongolia. 2012. Photograph by Scott L. Gardner. Used with permission.

Plate 14. Lacustrine vole *Microtus limnophilus* from near Har Us Nuur in the Shargyn Gobi of Mongolia. The team recovered the tapeworm *Echinococcus multilocularis* for the first time as larvae from these voles in Mongolia. 2012. Photograph by Scott L. Gardner. Used with permission.

MVPP-2012
NK-223782

Plate 15. Hundreds of *Echinococcus multilocularis* tapeworm larvae, floating inside each of the whitish balloon-like cysts, have replaced much of the liver of this lacustrine vole, *Microtus limnophilus*. This was the first record of this species in an intermediate host from Mongolia. In this parasite, a single egg may produce thousands of individual tapeworm larvae, each capable of developing into a 4 mm adult tapeworm. A single wolf host may harbor tens of thousands of individuals of this parasite species. 2012. Photograph by Scott L. Gardner. Used with permission.

Plate 16. This ural field mouse, *Apodemus uralensis*, taken from the steppe region near Takhi Station, southwestern Mongolia, harbored several species of helminth parasites. 2011. Photograph by Scott L. Gardner. Used with permission.

Plate 17. The Mongolian gerbil, *Meriones unguiculatus,* from Shargyn Gobi, Mongolia, is an intermediate host of the tapeworm *Taenia krepkogorski.* 2012. Photograph by Scott L. Gardner. Used with permission.

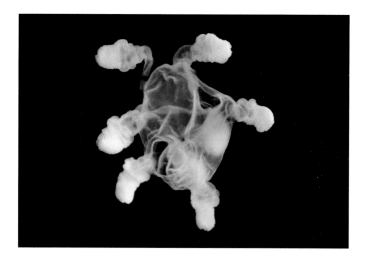

Plate 18. This *Taenia krepkogorski* tapeworm larva was recovered from a Mongolian gerbil, *Meriones unguiculatus*, from Shargyn Gobi, Mongolia. The gerbil ingested a single tapeworm egg from a fox or wolf feces. This larva developed in the gerbil's abdominal cavity; the image shows eight separate tapeworms hooked onto a central larval mass, all of which derived from the single egg. 2012. Photograph by Scott L. Gardner. Used with permission.

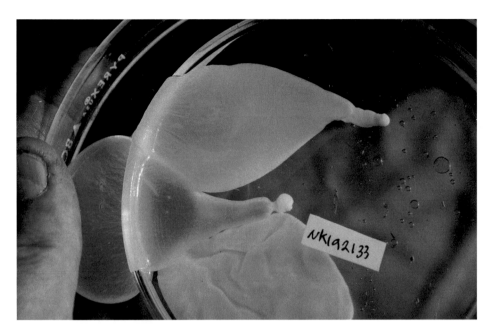

Plate 19. These *Taenia hydatigena* tapeworm larvae were recovered from a goat in the Gobi Desert of Mongolia. These tapeworms' definitive hosts in Mongolia are dogs and wolves, which both feed on goats. Each ingested egg develops into a single larva, called a bladder worm. 2010. Photograph by Scott L. Gardner. Used with permission.

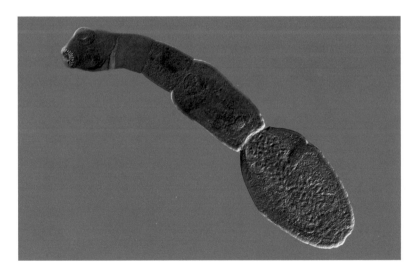

Plate 20. This *Echinococcus multilocularis* tapeworm was recovered from a fox on St. Lawrence Island by Robert L. Rausch during the 1950s. This species of tapeworm infected dogs which then transmitted the larvae to people in villages on the island. 2018. Photograph by Scott L. Gardner. Used with permission.

Plate 21. Sled dogs at the village of Gambell on St. Lawrence Island became infected by the tapeworm *Echinococcus multilocularis* and transmitted the parasite to people. 1954. Photograph by Robert L. Rausch. Used with permission.

Plate 22. Rice rats on Santiago Island, Galápagos, live in lava holes beneath *Opuntia* cactus trees. 2008. Photograph by Robert C. Dowler. Used with permission.

Plate 23. This rice rat, *Nesoryzomys swarthi*, from Santiago Island, Galápagos, is one of the few endemic mammals on the islands. This species is the definitive host of tapeworms of the genus *Raillietina*. 2008. Photograph by Robert C. Dowler. Used with permission.

Plate 24. This northern grasshopper mouse, *Onychomys leucogaster*, from Arapahoe Prairie, Nebraska Sandhills, harbored several individual *Hymenolepis robertrauschi* tapeworms. Photograph by Scott L. Gardner 2012. Used with permission.

Plate 25. Pocket gopher mounds are abundant near a branch of the Middle Loop River that cuts through the Nebraska Sandhills. The mounds show the extent of movements of soil made by Sandhills pocket gophers. 2012. Photograph by The Platte Basin Timelapse project. Used with permission.

Plate 26. Pocket gophers, *Geomys lutescens,* such as this one from Haythorn Ranch, Nebraska Sandhills, are favorite subjects for parasite studies. 2012. Photograph by Scott L. Gardner. Used with permission.

Plate 27. Site of tuco-tuco research at Estancia Agua Rica, near Mt. Sajama, in west-central Bolivia. 1986. Photograph by Scott L. Gardner. Used with permission.

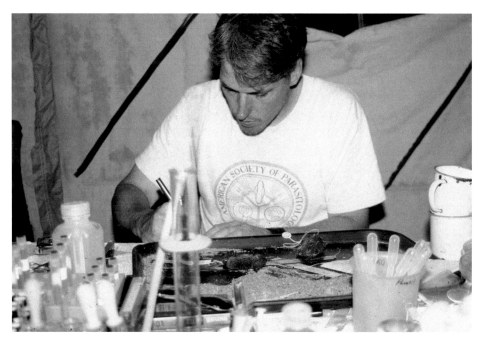

Plate 28. Scott L. Gardner preparing specimens in his field lab during the survey of parasites of mammals of Bolivia. 1996. Photograph by Scott L. Gardner and Terry L. Yates. Used with permission.

Plate 29. This species of tuco-tuco, *Ctenomys conoveri,* from south-central Bolivia, has the largest body size of any member of this genus of rodents. Individual tuco-tucos can harbor multiple species of nematodes, tapeworms, lice, fleas, mites, ticks, and protists. 1985. Photograph by Scott L. Gardner. Used with permission.

tuco-tucos in Bolivia, including those that lived on the vast Altiplano. Gardner's rare parasitic worm turned out to be present in every tuco-tuco species. In fact, tuco-tucos harbor an entire zoo of parasites that include, in addition to the *Paraspidodera* nematodes, whipworms, filarioid nematodes, tapeworms, and *Eimeria*, a single-celled protozoan that infects the digestive system.

Gardner asked where the parasites had come from. Did they arrive with the tuco-tuco ancestors when they journeyed from Africa? Or had the parasites originally lived in a local animal and then switched hosts when conditions were opportune? Gardner could find no close relatives in Africa to the nematodes found in the tuco-tucos. He suspected the worms might initially have lived in South American marsupials or in the native placental mammals such as armadillos, sloths, or anteaters. Gardner speculated that opportunistic host switching had occurred, that the parasites had jumped from their original hosts to the tuco-tucos. If host switching from another South American mammal had occurred, then there should be another species whose parasites closely matched those of the tuco-tucos.

Support for parasite host switching eventually came from research on the parasites of marsupials, armadillos, and a strange little mammal that descended from 50-million-year-old South American ancestors. Looking like a piece of salmon sushi, the pink fairy armadillo carries a pink plate of armor on its back with soft fur on its underbelly. They are nocturnal, burrowing into tunnels in various habitats ranging from sand dunes to grasslands, and once down inside, they block the entrance hole with their armor to prevent intruders from above. But these walking sushi pieces were not prepared for intruders from below, where tuco-tucos may have encountered them. Tuco-tucos can be aggressively territorial and have sharp incisors, and the interactions can't have been pleasant. Both groups consume a lot of soil and can ingest each other's feces during their respective burrowing activities, potentially enabling their parasites to switch hosts. There is a great deal of overlap in the ancestral nematode parasite faunas among the long-term South American natives, such as the fairy armadillos, and the more recent newcomers, the tuco-tucos. This suggests that sometime in the past, while they shared burrows, their parasites switched hosts from native to relative newcomer.

Parasitology is a never-ending process of solving mysteries, and there are abundant unsolved puzzles when it comes to the parasites of tuco-tucos. Gardner and his colleagues have recently found *Paraspidodera* nematodes in pocket gophers in central Mexico, and there is no clue yet as to how they ended up there. Separated by more than 2,000 km, pocket gophers and tuco-tucos are unlikely to have ever come into contact. The northernmost range for tuco-tucos is Lake Titicaca in Bolivia and Peru, but pocket gophers range from Canada south to Colombia. Somehow the parasites moved from one host to the other, traversing the Andes in the process.

Most parasites find food, habitat, and mates the same way other animals do. It is a misconception to think that these organisms are simply freeloaders. Throughout evolution, parasites have become diverse—specialists that increasingly become generalists and vice versa—occurring in hosts in every habitat, on all continents including Antarctica, and in every ocean. Parasites are constantly evolving, expanding their range of suitable hosts, moving into new geographic regions, and tweaking ecological networks across the world.

Chapter Thirteen

A Balancing Act

Almost 50 years ago, a 12-year-old boy on a farm in the Willamette Valley of Oregon dissected a camas pocket gopher and found a tapeworm. He followed in the footsteps of his uncle, Robert L. Rausch, director of the Arctic Health Research Center and one of the most influential parasitologists in the world. A decade later the boy realized the tapeworm was a species new to science, and he named it after the river near where it was first found. The name, *Hymenolepis tualatinensis*, literally means "a *Hymenolepis* from the Tualatin." That boy eventually became curator of the Harold W. Manter Laboratory of Parasitology at the University of Nebraska State Museum, which holds one of the largest parasitology collections globally. When Scott Gardner returns to the farm that he still owns with his family, there are plenty of camas pocket gophers around, but the tapeworm is nowhere to be found. Parasites are never static entities—their diversity and abundance fluctuate with subtle modifications in ecosystems and particularly with changes brought about by human environmental mismanagement.

The world's climate is changing at an unprecedented rate, pressuring animals and plants to move to new areas or risk extinction. As potential host animals migrate and come into contact with new communities of organisms, their parasites come along with them. In these new environments, the parasites sometimes switch hosts to take up residence in a different species. Switching hosts is one way that parasites end up escaping the fate of a host species bound for extinction.

Daniel R. Brooks and Eric P. Hoberg articulated the complex relationships among parasites, evolution, ecology, and climate change.

Referred to as the Stockholm Paradigm, this unified perspective describes how climate change is altering the relationships among people, parasites, livestock, crops, insects, and wildlife. There are several data sources—knowledge of diversity, insight into past environments, and awareness of key biological processes—that aid in anticipating the future in a world of rapid change. In this future, emerging diseases will erupt at a much faster rate, and scientists and policy makers need ways not only to react to them, but also to predict and prepare for their likelihood.

Brooks and Hoberg advocate using various kinds of proactive risk management rather than waiting for the aftermath of an emerging crisis. In 2014, Brooks, Hoberg, Gardner, and their colleagues offered a road map for how to predict biological responses to future changes. Not surprisingly, parasites feature at the center of the map, because they connect to so many other players in ecosystems. The road map is called DAMA, referring to its elements: document, assess, monitor, and act.

Documenting biodiversity is a crucial first step to understanding global change. Scientists have been systematically describing species even well before the 1700s when the Swedish biologist, Carl Linnaeus, published his famous binomial system for naming and classifying organisms. Biologists have named about a million species on land and about a quarter million living in the ocean, but these numbers pale in comparison with estimates of how many species actually live on Earth. Some models hover around 9 million, while others go up to 100 million. This huge discrepancy has a lot to do with microorganisms and parasites. Bacteria, archaea, and various kinds of worms are hidden players overlooked in many estimates of global species diversity. There are likely more than half a million species of nematodes, and their abundance is staggering—thousands in a gram of soil. Some estimates suggest that about 40% of all living things are parasites.

Every new species described adds information to the storehouse of knowledge about living organisms and the impacts of environmental change. Scientists compile taxonomic inventories, composed of not only data, but also the specimens themselves housed in museums, so their linked genetic and morphological data remain accessible for future studies. The best data rely on the descriptions of natural history that have been collected for centuries: For parasites, data are accumulated on the host, too, and include where was the organism found? What are its habitat preferences? How is it transmitted between hosts? What parts of the host are they found in? It turns out that documentation in the field and lab, wherever organisms are found, is the first step to being able to predict what will happen when things change.

The natural historians of the seventeenth and eighteenth centuries appreciated the uniqueness of species, and their accounts tell us a great deal. In the mid-nineteenth century, Darwin and Wallace proposed a comprehensive theory to account for the evolution of all life on Earth, producing a paradigm shift in how natural history information would forever be considered. Each living organism was now known to be linked to others through common descent, and the patterns of ancestry shape the way each individual's life plays out. Related organisms have similar ways of interacting with each other and their environment, but their differences pinpoint the specific kinds of selective pressures that shape new responses. Comparisons and contrasts are the tools of natural history, teasing apart each similarity and difference to assemble patterns of change over time. The assessment feature of the DAMA protocol asks scientists to assemble the big picture for each species: not only the basics of how it lives in its own ecology, but also the ecology of the hosts, the other organisms in the ecosystem, and how they have adapted and evolved through time.

The early natural historians viewed their observations and discoveries as contributions to be catalogued and stored in perpetuity in museum collections. Each new organism was once considered a separate reflection of God's creation. When Darwin and Wallace's influence spread widely, and naturalists began to understand the mutability of species, they still viewed their findings in a static way: Each organism had its own fixed position in an ecosystem, and its behavior was considered a consistent reflection of its genetic makeup. But environments are always changing, and organisms consistently show a remarkable ability to adapt, not just over generations, but also within the lifetime of a single individual. Scientists have a modern lexicon to communicate the dynamic nature of living things: phenotypic plasticity, adaptive change, variable gene expression, condition-sensitive behavior repertoire, homeostatic behavior, host switching, and many more. And each dynamic adjustment of one organism then triggers changes in each of the other organisms it interacts with.

Communities can be fully understood only by using monitoring systems to track them through time and space. It is no longer sufficient to collect a parasite at one place and time. Modern science requires continuous broad geographical sampling. Tracking systems allow scientists to create networks of information enabling comparisons across geographic space and over time. DNA barcoding of vouchered museum specimens and collaborations of scientists across geopolitical boundaries help develop the kind of information networks that will alert scientists to the critical changes in one location that reverberate in others.

Almost 50 million dedicated bird observers regularly monitor avian species. In the Northern Hemisphere holiday season, vast numbers of people brave the cold to participate in the Christmas Bird Count. About 6,000 trained scientists are employed as ornithologists in the United States, and they rely heavily on the data provided by the birders who document their observations on digital databases, like eBird. As a result of the huge input by citizen scientists, the geographic movements of birds are some of the best known of any animal group. The data are dynamic and continuously updated so that when a bird like the Mississippi kite extends its range into Nebraska for the first time, everyone knows it right away. Parasites may impact human lives to a much greater degree than birds do, but there are fewer than 1,000 scientists in the United States who primarily identify as parasitologists. And counting parasites has yet to become a fad for citizen scientists.

Parasitologists scramble to identify species as quickly and accurately as they can. They connect to their peers in other countries, sharing data and specimens in order to collectively create the networks that will aid in predictive studies. When parasite-related outbreaks occur, parasitologists work with medical researchers and community health workers to develop programs for remediation. Many long-term eradication programs are run by international health agencies in cooperation with local governments and organizations, and these have made tremendous progress at controlling African river blindness, schistosomiasis, and infections from hookworm, *Ascaris*, and whipworm. And yet, people are constantly being forced from their communities to live in conditions ripe for accelerated parasite adaptations, like host switching or immune resistance. Hosts, parasites, and the environment are in constant flux, so as one changes, the interactions of the others change too. How much information will be needed to attempt to mitigate the impacts of climate change on biological communities? Data acquisition continues to demand a high degree of integration so that it is networked across ecosystems, linking evolutionary and ecological patterns and connecting knowledge of the past, present, and future.

The message of the Stockholm Paradigm is that basic research programs must be linked with applications focused on how, in the long run, people and parasites are going to live together. In the assessment of biodiversity, parasites play a special role. Since parasites often rely on multiple host species to complete their life cycles, finding a parasite in just one host in an area can indicate healthy populations of that parasite's other host species. Parasites thus track an ecosystem's health and biodiversity, and their presence or absence can serve as environmental signals of the rates of change of other species.

Parasites are the bad kids on the block, inevitably portrayed as seductive, but always causing harm. But in many parasite-host relationships, the parasite rarely causes significant harm to the health of the host, because when the host dies, the parasite usually does too. In this way, parasite-host relationships often look very much like commensals—one partner in the relationship benefits, while the other neither benefits nor is significantly harmed. Some parasites have long life spans, and individuals may live more than 20 years in their host. All relationships among living things likely involve some degree of evolved dependence that results in at least one partner sacrificing some resources.

In the chaotic, unpredictable world of changing environmental conditions, parasites may help hosts adapt. Parasites may stimulate host immune systems to ward off new microbes, and they help hosts metabolize new forms of food. Tough times are the new normal for higher organisms on Earth. The resources that hosts sacrifice to their cohabitating parasites may be relatively insignificant compared with the benefits afforded by the partnership.

The planet is losing species faster than scientists can name them—much like burning a library without knowing the names or the contents of the books. In 1986 an arsonist set fire to the Los Angeles Public Library, resulting in the worst library fire in United States history. More than a million books were destroyed along with microfilm, patents, photographs, and magazines. It was as if a slice of human culture had been sucked out by an alien force. In the aftermath, thousands of volunteers assisted in the cleanup, restoring what could be saved and raising funds for replacements. Today more than 20 million species are now at risk of disappearing due to human activities, resulting from deforestation, use of toxic chemicals, habitat destruction, and global warming. Where the library's catalog could inform what was lost, our knowledge of the Earth's biodiversity catalog is still so incomplete that many species will vanish without ever having been identified. This is particularly true of parasites, since only a small fraction of them have been described. The imminent loss of parasite diversity will forever curtail our understanding of how entire communities of organisms interact and evolve.

Acknowledgments

We are indebted to the many researchers and students who have supported the work of the Harold W. Manter Laboratory of Parasitology. We particularly thank Professor Batsaikhan Nyamsuren and Professor Ganzorig Sumiya at National University of Mongolia and Altangerel (Auggie) Tsogtsaikhan Dursahinhan. Eric Hoberg provided valuable insights from his many years as chief curator of the former United States National Parasite Collection. Robert Dowler from Angelo State University graciously provided specimens and photographs for the Galápagos work. William C. Campbell reviewed our description of the discovery of ivermectin and gave us valuable insights. We are grateful for the encouragement given to us at the University of Nebraska and wish to particularly acknowledge the help of Amy Spiegel, Methodology and Evaluation Research Core, Paul Royster, University Libraries, and Aaron Sutherlen, professor of graphic design, who put us in touch with illustrator Brenda Lee, without whom this book would not have been possible. We remain appreciative to Alan Bond, professor emeritus of Biological Sciences, and to Sue Ann Gardner, professor of Libraries, who provided editorial assistance throughout the course of this project. Finally, we owe much to our editor at Princeton University Press, Robert Kirk.

This work was supported in part by the National Science Foundation through awards: BSR-8612329 *Zoogeography and Coevolution of Helminth Parasites and Their Rodent Hosts in Bolivia*; BSR-9024816 and DEB-9496263 *Parasites of Mammals of Bolivia: Phylogeny and Coevolution*; DBI-1756397 *Natural History: Digitizing and Conserving*

Specimens in the Manter Laboratory of Parasitology; DBI-145839 *Natural History: Securing and Digitizing Data for Parasite Biodiversity Specimens in the Manter Laboratory.* Any opinions, findings, and conclusions or recommendations expressed in this material are those of the authors and do not necessarily reflect the views of the National Science Foundation.

Appendix

A Guide to Parasites Mentioned

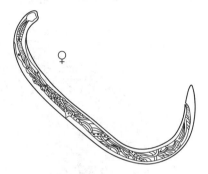

Figure G.1. *Ancylostoma duodenale*. Illustration by Brenda Lee.

Ancylostoma duodenale (female) **Human hookworm. Nemata. Size: 8–13 mm. Host: humans**. Adult worms attach to the intestinal villi of their human host. They scrape the villi and feed on the blood. Females lay thousands of eggs per day that pass out in the feces. Outside the host, the juveniles hatch in the soil, where they molt to the infective stage. They remain in the fecal matter in the soil until they encounter a human host. Once they contact a human, the juveniles penetrate the skin and travel via the bloodstream to the lungs. They migrate up the trachea and are then swallowed; when they reach the intestine, they molt to the adult stages and attach to the intestinal villi. Hookworm infection is common in tropical regions and frost-free temperate zones affecting hundreds of millions of people. Heavy infection causes daily blood loss, anemia, and other medical problems.

Anisakis simplex: **Herring worm. Nemata. Size: about 4 cm long, 1 mm wide. Intermediate host: 1st–zooplankton, 2nd–fish or squid. Definitive host: marine mammals**. In marine mammals, the adult worms attach to the lining of the stomach where they mate and produce eggs which pass out in feces in the ocean. *Anisakis* is one of the most common accidental parasites in humans, who become infected when eating raw or undercooked fish. When the worms attach to the stomach, they cause painful ulcers.

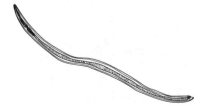

Figure G.2. *Anisakis simplex*. Illustration by Brenda Lee.

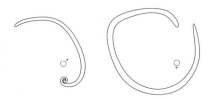

Figure G.3. *Ascaris lumbricoides.*
Illustration by Brenda Lee.

Ascaris lumbricoides. **Large human nematode. Nemata. Size: 15–50 cm long, 2–5 mm wide. Host: humans**. In the human intestine, female worms produce eggs which are passed in feces. People become infected by consuming contaminated food or water. *Ascaris* is common in tropical and some temperate regions, and more than 2 billion people are infected. Some infected people are asymptomatic, while others have mild discomfort. Children infected with numerous *Ascaris* exhibit serious problems that can turn deadly. In areas lacking adequate sanitation, infections are common.

Coitocaecum parvum. **Fish flatworm. Platyhelminthes. Size: 0.1–1 mm. Intermediate hosts: New Zealand mud snails and amphipods. Definitive host: common bullies,** *Gobiomorphus cotidianus*.
This species occurs in New Zealand. The infective stage, the miracidium, hatches in water and swims to a snail where it penetrates and moves to the digestive gland, where it multiplies asexually. Eventually the parasites emerge as cercariae, and they infect small amphipod crustaceans. The crustaceans are eaten by fish, where the worms travel to the intestine, mature, and produce eggs. Research on this species has shown that this worm can switch between alternative larval development strategies in response to changing environmental conditions and the availability of certain host species.

Figure G.4. *Coitocaecum parvum.*
Illustration by Brenda Lee.

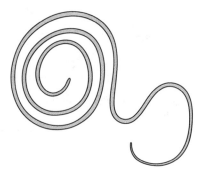

Figure G.5. *Crassicauda boopis.*
Illustration by Brenda Lee.

Crassicauda boopis: **Whale nematode. Nemata. Size: 1.5–2 m in length, 1–2 mm in width. Intermediate host: possibly crustaceans. Definitive host: fin whales, blue whales, and humpback whales**. This nematode occurs in the kidneys of large baleen whales. Scientific information about this species is limited, but some data show that transmission from mother to fetus can occur. Its life cycle is unknown, but it may use crustaceans as intermediate hosts. Necropsies on whales that died from natural causes show that parasitism with this large nematode causes kidney disease which can be fatal.

Figure G.6. *Cuscata* sp. Illustration by Brenda Lee.

Cuscata sp. **Strangle weed, dodder. Anthophyta. Size: 10 cm–30 m in spread, composed of 1–3 mm thick stems. Host: herbaceous and woody flowering plants**. This is a parasitic plant that lives on herbaceous plants, bushes, and trees. It absorbs water, minerals, and nutrients from its host plant using rootlike structures called haustoria. As the parasite grows, it gradually loses its ability to photosynthesize. Dodder does not have leaves, and the stems are yellowish in color due to the lack of chlorophyll. It has small flowers, and while many of its seeds are dispersed close to the mother plant, a few are ingested by birds and dispersed long distances.

Dicrocoelium dendriticum. **Lancet liver fluke. Platyhelminthes. Size: 5–10 mm. Intermediate host: 1st–land snails, 2nd–carpenter ants. Definitive host: mostly herbivorous mammals such as cattle and sheep**. This is a common parasite of wild and domestic ruminants throughout the world. Adult worms live in bile ducts of the liver, eggs pass out with bile into the intestine, and they are expelled with feces. When an egg is eaten by a land snail, the miracidia hatch and penetrate the snail's digestive gland, where it multiplies asexually, eventually producing thousands of cercariae. The cercariae exit the snail's mantle covered in slime that forms into balls. Carpenter ants are attracted to the slime balls, eat them, and become infected. While some cercariae encyst in the body cavity, others travel in the nervous system, turning the ants into zombies. Infected ants do not return to the colony but instead climb to the tops of grass blades where they attach for the night. Cattle and sheep eat the ants when they munch on grass. Once an infected ant is eaten, the parasites break out, travel to the liver, mature into adults, and begin producing eggs.

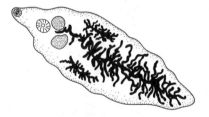

Figure G.7. *Dicrocoelium dendriticum.* Illustration by Brenda Lee.

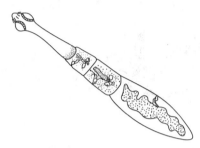

Figure G.8. *Echinococcus multilocularis.*
Illustration by Brenda Lee.

Echinococcus multilocularis. **Fox tapeworm.
Platyhelminthes. Size: 2–4 mm. Intermediate host:
voles, lemmings, mice. Definitive host: wolves,
coyotes, foxes**. Tapeworm eggs are passed in canid feces,
where they are accidentally ingested by rodents. In the
rodent, larvae penetrate the intestine and enter the
bloodstream, usually getting stuck in the liver, where they
form a continuously growing cyst. When a canid eats an
infected rodent, it acquires hundreds of tapeworms.
Humans usually become infected when a child plays
with a dog that has the tapeworms. When a tapeworm
egg is ingested by a person, it follows the route of blood
to liver, but the larvae can develop in almost any body
organ. Infection can go unnoticed for many years as the
cysts grow slowly. The disease is becoming prevalent in
Europe in areas where fox populations are increasing.

Eimeria. **Apicomplexa. Protista. Size: 15–30 µm. Definitive host:
Any vertebrate.** *Eimeria* is a genus of single-celled parasites with
thousands of species. In vertebrates, the parasites live in the lining
of the small intestine, where they first reproduce asexually. Later they
have sex and produce oocysts that are passed out in the feces. In the
feces outside the host, the oocysts sporulate into four infectious
sporozoites. When contaminated food is eaten, the oocysts are
breached, and sporozoites are released in the intestine of a new host.

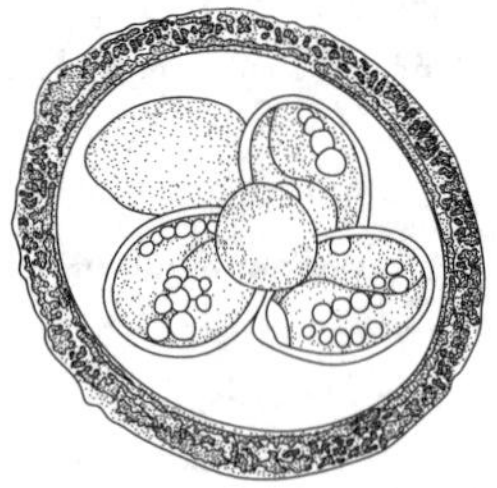

Figure G.9. *Eimeria* sp.
Illustration by Scott
Gardner.

Figure G.10. *Enterobius vermicularis.*
Illustration by Brenda Lee.

Enterobius vermicularis. **Human pinworm. Nemata.
Size: 3–13 mm. Host: humans.** Pinworms are one of
the most common parasites of people, infecting about
a quarter of the population living in temperate climate
zones at any one time. Pinworms live in the large intestine
of the host, where they feed primarily on bacteria and
intestinal epithelial cells. Females often migrate out the
anus and lay eggs on the surrounding skin. The eggs
cause itchiness, so when people scratch themselves, eggs
are picked up on their hands and can be dispersed easily
in the environment. Once pinworms are established in
households with children, eliminating them requires
thorough cleaning.

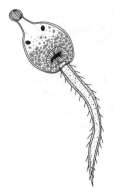

Euhaplorchis californiensis. **Trematode. Platyhelminthes. Size (adult): 0.25 mm in length, 0.1 mm in width. Intermediate host: 1st–horn snail, 2nd–killifish. Definitive host: fish-eating birds**. These parasitic worms live in the intestine of many shorebirds that feed in the marshy swamps and estuaries of Southern California. The larval trematode leaves the first intermediate host, a snail, and swims around using its simple eye to find its next host, a killifish. Cercariae penetrate the fish's skin and some travel to the brain. In the process, they change the fish's behavior to make it more vulnerable to bird predation. The worms then develop into adults in the birds' intestines.

Figure G.11. *Euhaplorchis californiensis.* Illustration by Brenda Lee.

Giardia duodenalis. **Giardia. Protista. Size: 12–15 μm. Host: mammals**. Infection with this parasite results when a potential host swallows resistant giardia cysts with contaminated water or food. Giardia cysts can survive in a moist environment for a long time. After ingestion, the cyst turns into a trophozoite that lives in the intestine of the host. These cells multiply via binary fission and eventually form more cysts that are dispersed in feces. Most infections happen when people drink contaminated water.

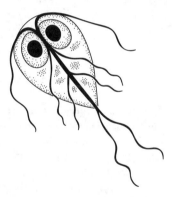

Figure G.12. *Giardia duodenalis.* Illustration by Brenda Lee.

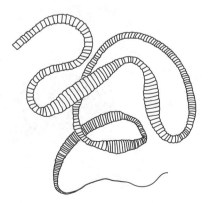

Figure G.13. *Hymenolepis diminuta.* Illustration by Brenda Lee.

Hymenolepis diminuta. **Rat tapeworm. Platyhelminthes. Size: 5–60 cm in length. Intermediate host: beetles (mealworms). Usual definitive host: rats of the genus *Rattus***. Beetles become infected when they eat rat feces with tapeworm eggs. Eggs hatch in a beetle and develop into the infective form, a cysticercoid. When a rat eats the beetle, the cysticercoid transforms into an adult in the small intestine. The tapeworm grows hundreds of proglottids that disperse eggs when they break off from the posterior end.

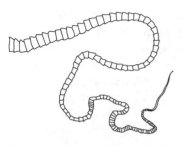

Figure G.14. *Hymenolepis robertraus-chi*. Illustration by Brenda Lee.

***Hymenolepis robertrauschi*. Grasshopper mouse tapeworm. Platyhelminthes. Size: 4–8 cm in length. Intermediate host: unknown beetles. Definitive host: grasshopper mice**. This species is a distant relative of the rat tapeworm, *Hymenolepis diminuta*. This tapeworm occurs in grasshopper mice from Nebraska to the Rio Grande Valley in New Mexico. The life cycle has been reproduced in mealworm beetles and in deer mice in a laboratory setting. No human infection is known for this tapeworm species.

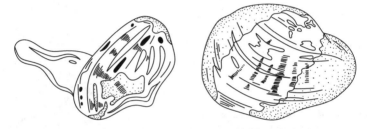

Figure G.15. *Lampsilis siliquoidea*. Illustration by Brenda Lee.

***Lampsilis siliquoidea*. Fatmucket mussel. Mollusca. Size: 7–11 cm (adult), 0.25 mm (larva).** The fatmucket mussel is common in northern freshwater rivers from the East Coast to the Rocky Mountains. Like many other freshwater mussels, it uses fish to disperse its offspring. The tiny larvae, called glochidia, attach to the gills of fishes. The gills often form scar tissue around the larvae, which can reduce the fish's ability to absorb oxygen from the water. In a few months, the larval mussel develops into a juvenile, drops off the fish, and begins an independent life.

***Leishmania*. Euglenozoa. Protista. Size: 1–10 μm. Intermediate host: phlebotomine sand flies. Definitive host: mammals**. *Leishmania* is a large group of single-celled parasites of mammals that live in tropical and subtropical regions. Infected mammals can be asymptomatic but still transmit the parasite when bitten by a sand fly. The parasites multiply in the sand fly, and when the sand fly bites another mammal, it transmits the protozoan. The disease caused by this parasite in humans has many names: kala-azar, oriental sore, or leishmaniasis. Symptoms in humans range from painless sores to life-threatening damage to internal organs.

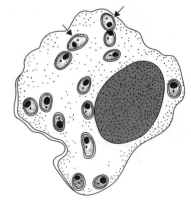

Figure G.16. *Leishmania* spp. Illustration by Scott Gardner.

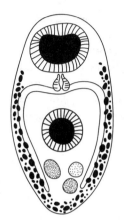

Leucochloridium variae. **Trematode. Platyhelminthes. Size: 1.3–2 mm. Intermediate host: land snails. Definitive host: songbirds**. These worms live and mate in the intestine of their bird hosts. Eggs are spread in the environment via bird feces. The eggs hatch after they are ingested by a snail. The miracidia travel to the digestive gland, turn into sporocysts, and eventually into cercariae which migrate into the eyestalks of a snail. There they form a sack full of individual worms, giving the eyestalks the appearance of large pulsating caterpillars, attracting the attention of birds. Once the snail is eaten by a bird, the larvae develop into adults in the bird's intestine.

Figure G.17. *Leucochloridium variae*. Illustration by Brenda Lee.

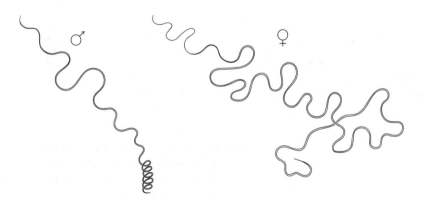

Figure G.18. *Litomosoides* sp. Illustration by Brenda Lee.

Litomosoides. **Nematode. Nemata. Size: 15–25 mm (males); 50–120 mm (females). Width: 0.1 mm. Intermediate host: mites. Definitive hosts: New World rodents, bats, and marsupials**. These long, slender worms live in the abdominal and pleural cavities of their mammal hosts. After mating, females produce microfilariae that disperse throughout the circulatory system. Parasitic mites pick up the microfilariae while feeding and act as vectors when they move to other hosts. This is an example of an endoparasite that utilizes an ectoparasite to infect new hosts.

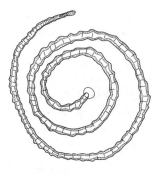

Moniliformis moniliformis. **Thorny-headed worm. Acantho-cephala. Size: 5 cm (males) up to 30 cm (females). Intermediate host: beetles and cockroaches. Definitive host: rodents and other small mammals.** Adult acanthocephalans live in the small intestine of their hosts. Males and females mate and their eggs pass in the hosts' feces. Cockroaches and beetles consume infected feces; in this host, the larvae develop into their infective stages, called cystacanths. When a mammal eats an infected insect, the worms are transferred and develop into adults in the small intestine.

Figure G.19. *Moniliformis moniliformis.* Illustration by Brenda Lee.

Myxobolus cerebralis. **Myxozoa. Cnidaria. Size: 7.5–400 μm. Intermediate host: annelids. Definitive host: fishes.** These parasites, distant relatives of jellyfish and corals, cause fish to swim in circles, giving the name, whirling disease. In salmon and trout, they invade the central nervous as well as other tissues. Spores are released when the host is eaten by larger fish or fish-eating birds. The spores are ingested by mud-dwelling tubifex worms, where the parasites develop in the worm's gut into the infectious stage. They are passed out in feces into water where they infect new fish hosts.

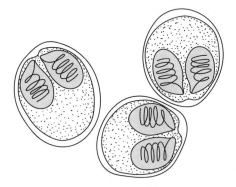

Figure G.20. *Myxobolus cerebralis.* Illustration by Brenda Lee.

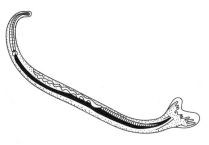

Figure G.21. *Necator americanus.* Illustration by Brenda Lee.

Necator americanus. **New World hookworm. Nemata. Size: 5–14 mm in length. Host: humans.** Hookworms have large mouths equipped with sharp teeth that cut into intestinal tissue. Because of the damage to the inner surface of the intestine, heavy infection with this small nematode can cause significant blood loss and anemia. Adult male and female worms mate in the small intestine of the host, and a single female can produce up to 9,000 eggs a day. Eggs are expelled in feces, and the juveniles live in the feces-laden soil. When a person contacts the soil, the juveniles penetrate the skin and migrate to the lungs, where they are coughed up and swallowed. In the small intestine, they mature into adults and remain for years.

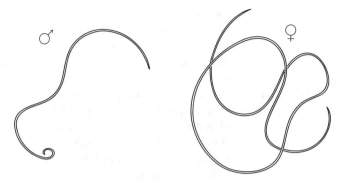

Figure G.22. *Onchocerca volvulus*. Illustration by Brenda Lee.

***Onchocerca volvulus*. River blindness worm. Nemata. Size: 23–70 cm (females), 1.5–5 cm (males). Intermediate host: blackflies. Definitive host: humans.** Adults of this filarioid nematode live in fibrous nodules in the skin of their human hosts. Eggs hatch into microfilariae that circulate in the skin. When the vector, a female blackfly, takes a blood meal, the microfilariae are ingested, go through three molts, and migrate to the salivary glands. When the blackfly bites another person, the microfilariae pass via the saliva into the host. The worms produce a range of skin lesions, but severe effects occur when the microfilariae cause inflammatory damage to the eyes, eventually resulting in blindness.

Figure G.23. *Orthohantavirus* sp. Illustration by Brenda Lee.

***Orthohantavirus*. Hantavirus. Size: 100–300 nm. This is a large group of RNA viruses that infect rodents, bats, shrews, and moles.** In the process of replication, the virus takes a small part of the host cell membrane, enveloping the virus particles. The virus is normally transmitted in feces or urine of an infected mouse. These viruses usually have little impact on their natural host. Human infections are generally traced back to contact with infected rodents. North American hantaviruses discovered in the 1990s cause hantavirus pulmonary syndrome, a deadly zoonotic disease in humans.

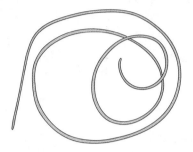

Figure G.24. *Paragordius tricuspidatus.* Illustration by Brenda Lee.

Paragordius tricuspidatus. **Old World horsehair worm. Nematomorpha. Size: 120–310 mm. Width: 0.09–0.4 mm. Hosts: aquatic invertebrates and terrestrial crickets and grasshoppers. Definitive host: none, adults are free-living.** The adult worms live their short lives in streams and freshwater ponds. Also known as Gordian worms, the adults look like an entangled ball of thread when they come together to mate. They lay eggs on sticks in water; when the eggs are ingested by larger invertebrates such as snails or insects, they hatch and penetrate the gut of the host and turn into cysts. When the water recedes, the hosts die and are eaten by crickets or grasshoppers. Larvae penetrate the gut of the new host and grow in the body cavity to length of a shoelace. This worm does not kill the cricket but can occupy a large part of its body. Once fully developed inside the host, the worm affects the nervous system of the cricket to cause it to jump into water. Within minutes the long filamentous adult worm emerges from the cricket.

Paraspidodera **sp. Tuco-tuco worm. Nemata. Size: 10–28 mm in length. Intermediate host: none. Definitive host: tuco-tucos.** This parasitic nematode occurs in rodents that are endemic to South America and Mexico, such as tuco-tucos, agoutis, pacas, guinea pigs, and pocket gophers. Nematodes living in the rodent's cecum mate and shed eggs in their host's feces. Other individual rodents sharing the burrow system accidentally ingest the nematode eggs. Scientific research on this parasite gives clues to the evolution of rodents and their parasites in South America.

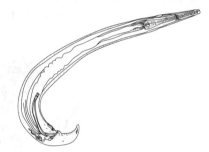

Figure G.25. *Paraspidodera uncinata.* Illustration by Scott Gardner.

Figure G.26. *Placentonema gigantissima.* Illustration by Brenda Lee.

Placentonema gigantissima: **Giant whale nematode. Nemata. Size: 2–8 m in length. Width: 2.5 cm. Intermediate host: unknown. Definitive host: sperm whales.** This nematode lives in the placenta, uterus, and mammary glands of female sperm whales. The only known specimens were collected in the early years of California whaling stations. It is considered one of the largest nematodes in the world, but there is very little scientific information about the biology of this worm.

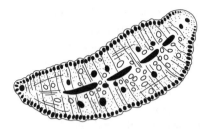

Figure G.27. *Placobdelloides jaegerskioeldi.* Illustration by Brenda Lee.

***Placobdelloides jaegerskioeldi.* Hippo butt leech. Annelida. Size: 27–30 mm. Host: hippopotamus**. This unusual leech lives in the anal folds of hippopotamuses. These leeches have a simple life history: they are hermaphrodites, but individuals mate with each other and produce eggs. Eggs are laid in cocoons attached to hard surfaces. Juvenile leeches are similar to adults, and there are no larval stages. After feeding, most leeches detach from their host and live in stream bottoms, often not feeding again for months.

***Plagiorhynchus cylindraceus.* Thorny-headed worm. Acanthocephala. Size: 4–13 mm. Intermediate host: soil-dwelling isopods. Definitive host: songbirds**. These worms live and mate in the gut of songbirds, such as starlings or robins. Eggs excreted in the feces of the bird are eaten by isopods. The eggs hatch in the isopod and penetrate the body cavity, where they turn into cystacanth larvae. The presence of the larvae can increase the isopod's foraging activities and decrease its light avoidance, making it more likely to be found and eaten by birds. When infected isopods are eaten by songbirds, the larvae develop into adult worms in the intestine.

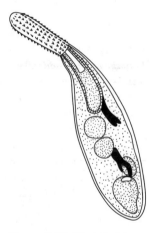

Figure G.28. *Plagiorhynchus cylindraceus.* Illustration by Brenda Lee.

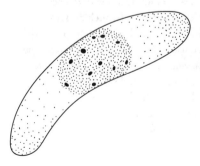

Figure G.29. *Plasmodium falciparum.* Illustration by Brenda Lee.

***Plasmodium falciparum.* Apicomplexa. Protista. Size: 1–20 μm. Intermediate host: primates. Definitive host: *Anopheles* mosquito**. The parasite lives in mosquitoes, sexually reproducing in the lining of their gut. After fertilization, the oocyte releases sporozoites, which migrate to the mosquito's salivary glands. When the insect takes a blood meal from a primate, the parasites are injected along with saliva into the host's blood. The injected parasites first multiply in the host's liver cells and later invade red blood cells. *P. falciparum* is the most deadly species of the *Plasmodium* parasites. It occurs in tropical regions, causing malignant tertian malaria, most often in young children.

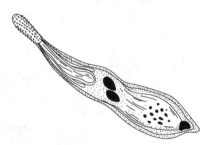

Figure G.30. *Polymorphus minutus.*
Illustration by Brenda Lee.

Polymorphus minutus. **Thorny-headed worm.
Acanthocephala. Size: 3–12 mm. Intermediate host:
freshwater amphipods. Definitive host: water birds.**
In the small intestine of their avian hosts, these
acanthocephalans mate and produce eggs. The eggs
are dispersed in feces and are ingested by amphipods,
where the eggs hatch and develop into larval cystacanths. The parasite changes the host's behavior by
making it attracted to light. Infected amphipods swim
toward the surface of the water and cling to material
there, making them more susceptible to be eaten by
water birds. Once inside a bird, the larvae emerge,
attach to the intestine, and develop into adults.

Protospirura ascaroidea. **Nematode. Nemata.
Size: 3–9 mm long, 0.4 mm wide. Intermediate
host: insects. Definitive host: rodents.** These
nematodes occur in the stomachs of many species
of rodents. Eggs are dispersed in feces where they
are consumed by insects, where the juvenile
nematodes encyst. When a rodent consumes an
infected insect, the worms develop into adults in
the stomach, where they mate and produce eggs.

Figure G.31. *Protospirura ascaroidea.* Illustration by Brenda Lee.

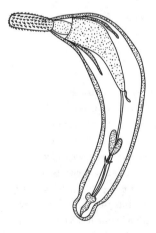

Figure G.32. *Pseudocorynosoma constrictum.* Illustration
by Brenda Lee.

Pseudocorynosoma constrictum. **Thorny-headed worm.
Acanthocephala. Size: 2.2–4.3 mm. Intermediate host:
amphipods. Definitive host: water birds, such as ducks.**
Adults of these thorny-headed worms live in the intestines of
ducks. Female worms produce eggs that are expelled in feces and
then eaten by amphipods, where the eggs hatch and turn into
larvae called cystacanths. As the cystacanths develop, they turn
bright orange and are visible through the semitransparent cuticle
of the amphipod. When the crustacean is eaten by a bird, larvae
attach to the wall of the intestine and mature into adults.

Figure G.33. *Rafflesia* sp. Illustration by Brenda Lee.

***Rafflesia* spp. Corpse lily. Anthophyta. Size: 5 cm–1 m. Host: other angiosperms**. This parasitic plant occurs in Southeast Asia, mainly in Indonesia, Malaysia, and the Philippines. One species has the largest flowers known. It has no leaves but develops a rootlike structure (haustorium) that enters the stem or root of a host plant and takes up nutrients. The most notable feature of this plant is a flower with the smell and color of a decomposing carcass, which attracts flies and beetles that disperse the pollen. Their seeds are distributed in the feces of small mammals, like tree shrews, that feed on the fruit.

***Raillietina* spp. Tapeworm. Platyhelminthes. Size: 2–25 cm. Intermediate hosts: ants, beetles, and other arthropods. Definitive host: mammals and birds**. These tapeworms live in the intestines of wild and domestic mammals and birds. All of the more than 750 known species of *Raillietina* have small spines on the suckers and circles of hammerhead-shaped hooks on the anterior part of the scolex. The larval tapeworms infect insects, and when the insects are eaten by birds and mammals, adult tapeworms develop in their intestines.

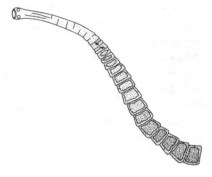

Figure G.34. *Raillietina* sp. Illustration by Brenda Lee.

Figure G.35. *Ransomus rodentorum*. Illustration by Brenda Lee.

***Ransomus rodentorum*. Gopher nematode. Nemata. Size: 8–9.2 mm long. Intermediate host: unknown. Definitive host: pocket gophers**. This nematode occurs only in pocket gophers, where adults live in the cecum and large intestine. Relatively few individuals live in a single gopher, usually no more than five or six. It is not known what limits their numerical density inside the hosts, whether it is the result of competition among the nematodes or the actions of the gopher's immune system.

Figure G.36. *Schistosoma mansoni*. Illustration by Brenda Lee.

Schistosoma mansoni. **Blood fluke. Platyhelminthes, Size: 10–20 mm long, 0.8–1 mm wide. Intermediate host:** *Biomphalaria* **snails. Definitive host: rodents and primates**. In humans, the adult schistosome worms live in the portal veins that drain the large intestine. Unlike other trematodes, these worms have separate sexes that live in permanent copula. Females lay hundreds of eggs per day that are passed into the blood. About two-thirds of the eggs remain in the host, trapped in the liver and other organs. The rest are expelled in feces. In water, the miracidia hatch and swim until they find a *Biomphalaria* snail. In the snail, they migrate to the digestive gland, multiply, and eventually produce cercariae with bifurcated tails. The cercariae break out of the snail and swim to the top of the water column. The cercariae penetrate soft parts of the skin when people, usually kids, wade or swim in the water.

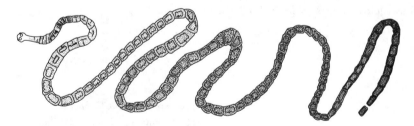

Figure G.37. *Taenia saginata*. Illustration by Brenda Lee.

Taenia saginata: **Beef tapeworm. Platyhelminthes. Size: 4–12 m long, 1 cm wide. Intermediate host: cattle. Definitive host: humans**. Adult *Taenia* worms produce tens of thousands of eggs per day in the small intestine of humans. Eggs passed in feces can contaminate grazing areas in regions with poor sanitation. When cattle feed on contaminated grass, they ingest the eggs, which then hatch, penetrate the intestine, and travel to muscle tissue. When a person eats raw or undercooked meat, they may ingest the infective tapeworm larvae, which then grow into adults in their intestine.

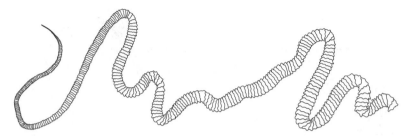

Figure G.38. *Tetragonoporus calyptocephalus.* Illustration by Brenda Lee.

Tetragonoporus calyptocephalus. **Whale tapeworm. Platyhelminthes. Size: Up to 30m. 1st intermediate host: marine crustaceans, 2nd intermediate host: marine fishes or squids. Definitive host: sperm whale.** This parasite is one of the longest tapeworms known. Little is known about its ecology, but related species have two intermediate hosts: the first are small planktonic crustaceans and the second are fish or squid that feed on them. When these hosts are eaten by sperm whales, the tapeworms mature in the small intestine to their colossal size. Fecal analysis of sperm whales has shown that this tapeworm is more common in some regions, suggesting differences in the diet of these top marine predators.

Toxoplasma gondii. **Apicomplexa. Protista. Size: 9–13 μm. Intermediate host: any vertebrate. Definitive host: cats.** This parasite lives in the intestinal cells of all cats, where they multiply and produce egglike oocysts in feces. The oocysts are extremely hardy and can remain viable for months. When a vertebrate—mouse, bird, dog, or human—ingests the oocyst, the parasite enters the bloodstream and forms cysts in various organs which multiply rapidly until they are slowed by the host's immune system. When a cat eats an infected vertebrate, the parasites break from their cysts, penetrate the intestine cells, and undergo sexual recombination which results in oocysts. The protist remains in its intermediate hosts forever, so when the host is later exposed to diseases like AIDS that weaken the immune system, the parasite grows rapidly and causes deadly illness. In pregnant women, the parasite can cross the placenta and infect the brain of the developing fetus.

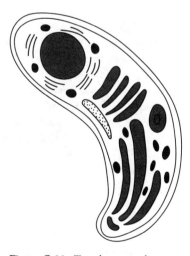

Figure G.39. *Toxoplasma gondii.* Illustration by Brenda Lee.

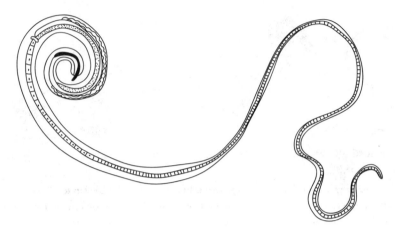

Figure G.40. *Trichuris trichiura*. Illustration by Brenda Lee.

Trichuris trichiura. **Whipworm. Nemata. Size: 3–5 cm. Host: humans**. These adult nematodes live in the large intestine, and females produce thousands of eggs per day that are passed in feces. If a person accidentally ingests an infective egg, the juvenile hatches in their intestine, then embeds its anterior end in the wall of the intestinal mucosa, where it grows to adult size. The nematodes can live there for many years. Whipworm infection is common in tropical and warm temperate climates.

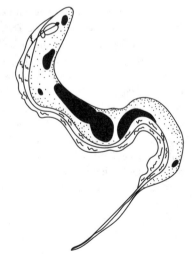

Trypanosoma cruzi. **New World trypanosome. Protista. Size: 16–20 μm. Host: mammals**. This parasite causes Chagas disease in the Americas. It infects mammals and prefers to hide in a host's muscle tissues, including cardiac muscle in the heart. The parasite becomes infective when it travels in the blood. When reduviid or kissing bugs feed on a vertebrate host, they suck in the parasite along with blood. Inside the bug, the parasite multiplies massively, and each trypanosome migrates to the bug's hind gut, where it remains until it is passed in the feces. As the bugs feed, they leave feces on the skin of mammals. The parasites enter the body via the bite opening or through mucus membranes, where they are ingested by white blood cells, multiply, and eventually travel to muscle and other tissues.

Figure G.41. *Trypanosoma cruzi*. Illustration by Brenda Lee.

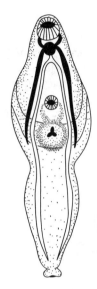

Uvulifer ambloplitis. **Common black spot. Platyhelminthes. Size: 1.3–2.3 mm. 1st intermediate host: snails. 2nd intermediate host: small fishes. Definitive host: fish-eating birds, especially kingfishers**. In birds, the adult worms expel their eggs into the intestine, where they are released via feces into water. The eggs hatch in water, where miracidia emerge, swim, and penetrate aquatic snails. In the snail, the parasite develops through stages and then multiplies, eventually producing cercariae. These emerge from the snail and swim to the top of the water where they contact a fish. Then the cercariae penetrate the skin of the fish, where they form a cyst that turns black. When an infected fish is eaten by a kingfisher, the parasites develop into adults.

Figure G.42. *Uvulifer ambloplitis*. Illustration by Brenda Lee.

Viscum album. **European mistletoe. Anthophyta. Size: 10–150 cm. Host: many kinds of trees**. The common mistletoe grows on trees and steals water and essential nutrients from its host plant. It is also capable of photosynthesizing its own nutrients. Mistletoe is dispersed to new hosts by its fruit, favored by birds. The seed is surrounded by a viscous, sticky flesh and often adheres to the bird's beak. The bird scrapes off the sticky stowaways as it travels, allowing the parasite seed to take root on a new host. The presence of mistletoe stresses the host tree, making it more susceptible to disease, other infections, and drought.

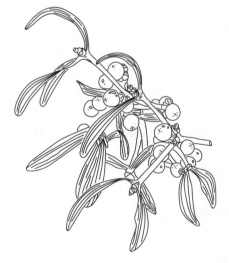

Figure G.43. *Viscum album*. Illustration by Brenda Lee.

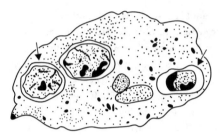

Figure G.44. *Wolbachia* sp. Illustration by Brenda Lee.

***Wolbachia* (shown inside a cell). Bacteria. α-Proteobacteria. Size: 0.8–1.5 μm. Hosts: arthropods and nematodes**. This group of gram-negative bacteria infects both arthropods (insects, spiders, crustaceans), and some nematodes. While some species of *Wolbachia* behave like parasites, other species have a symbiotic, mutually beneficial relationship. In some cases, the symbiosis is so strong that the survival of both parasite and host depend on the presence of each other.

Glossary

acanthocephalans are also known as thorny-headed worms. They are a little-known phylum of parasites, characterized by a club-shaped anterior attachment organ covered with tiny hooks. This proboscis anchors the worm to the intestine wall of its host. The thorny-headed worm's body is a tube that contains the reproductive organs but no mouth or gut, and it absorbs nutrients through its body surface.

acidification is the chemical process of water interacting with substances such as carbon dioxide and sulfur dioxide that lower the pH.

adaptive changes are modifications that enable organisms to become more likely to reproduce and produce offspring. Adaptive change can occur through behavioral, physiological, and/or evolutionary processes.

agoutis are large rodents related to porcupines and guinea pigs. They live in the tropics from Mexico to South America.

amoeba is a single-celled organism that moves by creating protrusions of its cell membrane.

amphibians are a class of vertebrates that includes toads, frogs, newts, and salamanders.

amphipods are a subgroup of crustaceans with laterally compressed bodies.

anthelmintics are a class of drugs used to treat many types of parasitic worm infections.

antibodies are protein molecules produced by the body as part of the immune system. Antibodies can recognize, bind, and neutralize antigens. Antibodies occur only in vertebrate animals.

antigens are substances recognized as foreign by the immune system of a vertebrate.

archaea is a group of ancient single-celled microorganisms believed to be ancestors of plants, animals, and fungi.

archipelagos are chains of islands close to each other that share a common geological origin.

arthropods are invertebrate animals with segmented legs and exoskeletons. The group includes insects, arachnids, crustaceans, and myriapods.

avermectins are a group of chemically similar antiparasitic drugs that include the first anthelmintic drug, ivermectin. This family of drugs is effective against nematodes and parasitic arthropods such as lice, mites, and bedbugs.

Ba'Aka is an ethnic group living in the rainforests of central Africa, primarily in the countries of Gabon, Cameroon, the Central African Republic, and the Republic of the Congo.

Bantus are Indigenous ethnic people of Sub-Saharan Africa who speak variants of the Bantu languages. More than 400 ethnic groups are in this language family.

Bering Strait is a narrow shallow seaway that separates the Pacific and Arctic oceans and lies between Russia and the state of Alaska in the United States.

binomial nomenclature refers to the convention of using two names to scientifically identify every distinct species on Earth. The first part is the genus name and the second is the species designation.

biodiversity refers to the variety and variability of living organisms in a geographic area and it includes their ecological relationships.

bioluminescence is the process in living organisms that uses organic molecules to convert chemical energy into light.

Carnivora refers to members of an order of predatory or scavenging mammals that share common ancestry. The group includes dogs, cats, bears, weasels, hyenas, civets, and mongooses.

castration is any process that inhibits or interferes with the production of ova or spermatozoa.

CDC refers to the Centers for Disease Control and Prevention, a federal agency under the United States Department of Health tasked with the research, monitoring, preparation, and prevention of infectious and other diseases.

cecum is a pouchlike structure at the junction of the small and large intestines in mammals.

cell membrane is a thin, flexible, double layer of lipid surrounding a cell and enclosing its plasma.

cell wall is a rigid shell immediately outside a cell membrane. Plants have cell walls made of cellulose, while bacteria have cell walls made of protein and sugar macromolecules.

cercariae are the ultimate developmental stage of a larval parasitic trematode. They often resemble a microscopic tadpole with a thin swimming tail attached to an immature but recognizable flatworm body.

cestodes or tapeworms are specialized parasitic flatworms. They have no mouth or gut and absorb nutrients through the protective surface covering their body, called the tegument. Most tapeworms are made up of small identical segments, called proglottids, that contain reproductive organs, and when full of eggs, they break off from the posterior end.

Chagas disease is caused by a parasitic infection of the protist *Trypanosoma cruzi*. It is commonly transmitted by reduviid "kissing" bugs, although eating contaminated food or juice can also cause an infection. The distribution of this parasite is limited to the Americas, and while the disease is not always fatal, chronic infection can lead to poor health and early death.

Christmas Bird Count was first organized by the Audubon Society in 1900 and has remained an annual event on Christmas Day when birds are counted throughout the Western Hemisphere. The survey is run by volunteers who record all observed bird species and their numbers, providing information on the abundance and distribution of avian populations in these regions.

chromosomes, components of cell nuclei, are made up primarily of DNA, the cell's genetic code.

Cichlidae is a large and diverse family of fish, common in tropical regions of Asia, Africa, South and Central America.

citizen scientists are nonprofessionals who collect research data in collaboration with scientists.

coccidia are single-celled parasitic protists that grow and multiply inside the cells of the intestinal wall of their host. These parasites reproduce both asexually and sexually.

coelacanths are an ancient group of lobe-finned fishes found in deep ocean waters off Africa and the Australasian region.

commensalism is an evolved relationship between two species in which one individual benefits, and the other has neither apparent positive or negative effects.

cryotubes are special plastic containers used to collect tissue samples and are often frozen in liquid nitrogen.

curators conduct research on and administer museum collections, exhibits, and education.

cysticercosis is a disease usually caused by larvae of *Taenia* tapeworms.

cysticercae are the larvae or intermediate stage of some tapeworms.

DAMA is a protocol or research framework for documenting, assessing, monitoring, and taking action with regards to changes in the ecology of parasitic organisms and other pathogens.

database is a digitally organized collection of information.

dire wolf or *Aenocyon dirus,* is a species of wolf that lived in what is now known as the Americas and went extinct around 10,000 years ago.

DNA, or deoxyribonucleic acid, is the molecule responsible for storing and transmitting genetic information in many living organisms.

eBird is a digital platform, ebird.org, that collects observations of bird sightings and scientific data on bird distributions, migrations, and abundance.

ecological fitting describes how some organisms can rapidly inhabit new environments.

ecology is the biological study of the interactions of living organisms and their environment.

ecosystem is a region that includes the living and nonliving components of the environment.

ectoparasites are organisms that usually live on the skin, scales, feathers, or shells of their host.

Eimeria is a genus of parasites common in vertebrates. They are protists and live inside cells of the intestinal epithelium.

elephantfish are members of the freshwater fish family Mormyridae. They are found in Africa and have a trunk-like fleshy extension that enables them to find prey in muddy river bottoms.

ELISA test detects specific antigens or antibodies in samples of blood serum. It is often used to detect signs of viral, bacterial, or parasitic infections in organisms.

endemic refers to organisms living in a specific region that have evolved there.

endoparasite is an organism that lives inside another species and often harms its host.

endosymbiont is an organism that lives inside a cell or body of another organism and may provide a benefit to its host.

Eocene is an epoch in Earth history that lasted from about 56 million years ago to about 34 million years ago.

eukaryote refers to the group of organisms that includes protists, fungi, plants, and animals. Their cells have a nucleus that encloses the genetic material of the cell.

family refers to a taxonomic group of genera that are closely related via a common ancestor.

filarioids are long, thin nematodes that live in tissues and are transmitted from host to host via biting arthropod vectors.

fluke is a common name for parasitic flatworms in the group Trematoda.

gastropod is a subgroup of molluscs that includes snails and slugs.

gene is a distinct sequence of DNA that codes for specific proteins or RNA.

glacial maximum marks times in Earth's history when the extent of ice cover on the continents was most extensive.

glochidia are tiny larval clams that are parasitic on the gills of fishes.

gram-negative bacteria are not stained by crystal violet, a diagnostic stain commonly used by researchers and physicians. The lack of staining indicates that these bacteria have a different kind of cell wall compared to gram-positive bacteria.

gram-positive bacteria are a group of bacteria that turn blue and pink from a common diagnostic stain, crystal violet. This staining method provides a rough identification of different bacteria.

hexacanths are larval tapeworms that emerge from the egg with six hooks that enable them to tear through the gut tissue of their intermediate host.

homeostatic behavior describes how living organisms make behavioral adjustments to function in a fluctuating environment.

host switching occurs when parasites move from one host species to a new host species.

hydatid cyst is a cyst that contains watery fluid and tiny larval *Echinococcus* tapeworms.

immune resistance occurs when a pathogen avoids detection by a host organism's immune system.

incidental host is an accidental host that usually results in a reproductive dead end for a parasite.

indigenous populations are people from ethnic communities distinguished by their culture, traditions, and histories, and who identify with geographic regions regarded as their ancestral lands.

ivermectin is a drug effective against some nematodes and parasitic arthropods such as mites, lice, and fleas.

jaundice is a condition of yellowish skin and eyes caused by inability of the liver to break down bilirubin.

karyotype is a preparation of stained cells on a microscope slide that can be used to show the number and structures of chromosomes.

King Leopold ruled Belgium between 1865 and 1909. He forced the occupation of much of the Congo basin in Africa and extracted its resources under brutal conditions.

kissing bugs are reduviid bugs, a group of true bugs that feed on blood.

mammoths are extinct giant mammals, related to modern elephants. They were covered in hair and lived in cold climates.

mange is a condition of severe hair loss caused by mites that burrow in the skin.

marsupials are mammals that give birth to relatively underdeveloped young that are nursed in a pouch.

mastodons are extinct relatives of elephants and mammoths. Mastodons have distinctive molars, suggesting they had different food preferences than elephants and mammoths.

Mesozoic, or the Age of Reptiles, is a period in Earth's history that lasted from 252 million to 66 million years ago and includes the Cretaceous, Jurassic, and Triassic eras.

metamorphosis is a process of change in form and shape from one developmental stage to the next, usually from one larval form to another or to an adult.

microfilariae are the microscopic larvae of certain parasitic filarioid nematodes.

microhabitat is the local environment where an organism lives.

miracidium is the first larval stage of a parasitic trematode after hatching from the egg.

mitosis is a process of cell division in which the descendant cells each receive a full set of chromosomes from the parent.

Mollusca is a phylum of invertebrates that includes snails, slugs, clams, mussels, squids, and octopuses.

monogenes are aquatic flatworms that parasitize fishes and amphibians.

morphology refers to the study of the form or physical appearance of an organism.

mutability is the tendency to change, particularly in reference to genetic material. Some components of DNA are more mutable than others.

mutation occurs when hereditary material, usually DNA or RNA, is modified, changing the encoded sequence.

mutualism describes an evolved relationship between two or more species that is beneficial for all of them.

naturalizing is the process by which a plant or animal species becomes established outside its usual range in a new geographic area.

nauplius is the first larval stage of crustaceans.

Nemata is the phylum name for nematodes.

niche is an ecological term that describes the role of a species within a biological community.

oncosphere or hexacanth refers to the six-hooked larval tapeworm that emerges from the egg.

oocyst is the infective and resistant form of protozoan parasites that passes out of the host.

organelle is a structure in a cell, such as mitochondria, endoplasmic reticulum, or nucleus, that performs a specific function.

pacas are large rodents in the genus *Cuniculus* that are native to forests of Central and South America. They have distinctive brown coats with white dots and stripes.

parasitism is an evolved relationship between two species in which one benefits from the interaction, and the other, the host, is negatively affected.

pentastomes or tongue worms are a group of parasitic crustaceans that infect vertebrates. They are most common in the tropics.

phenotypic plasticity refers to a mechanism by which genetically similar organisms can show dissimilar traits in response to different environmental conditions.

phylogeny describes the evolution of a group of organisms based on the history of their shared descent.

phylum is a primary taxonomic grouping of organisms with similar evolutionary histories.

placental mammals are mammals in which the embryo and the fetus develop in the mother's uterus and are fed and nourished through the placenta.

Plasmodium is a genus of parasitic protists that include several species that cause malaria.

platyhelminths or flatworms are an ancient group of animals with a simple shape and few organs. Some specialized parasitic flatworms, such as tapeworms, have no circulatory system or digestive tract, but only reproductive organs.

Pleistocene is a period in Earth's history that lasted from 2.6 million to 12,000 years ago, characterized by reoccurring ice ages.

plesiosaurs are extinct predatory marine reptiles that lived during the Mesozoic from about 200 million to 66 million years ago.

proglottids are the identical segments in tapeworms that act as independent replicating and egg-producing units.

pronghorn antelope (*Antilocapra americana*) is a species of native North American hoofed mammal. One of the world's fastest land animals, its closest living relatives are giraffes.

Protista or protozoa are a diverse group of single-celled or colonial eukaryotic organisms.

pulmonates are snails and slugs that breathe with a lunglike organ that allows them to live on land.

RNA, or ribonucleic acid, is a macromolecule similar to double stranded DNA, but with only a single strand. Some viruses use RNA as primary genetic material. Most organisms use RNA in protein synthesis.

schistosomes or blood flukes are a group of parasitic trematode worms that live in the circulatory system of their host.

Simulium is the genus that includes the blackfly, a vector for the nematode worm that causes river blindness. The larvae live in fast-flowing fresh water, and adult females feed on blood.

slave economy is a term for a commercial system that is built on institutionalized slavery and on the exploitation of slave labor.

species diversity is a measurement of the number of different species in a particular area weighted by the abundance of each species.

species richness is the number of different species occurring in a given habitat or geographic area.

sporocyst is a developmental stage of a parasitic trematode that develops inside the tissues of a mollusc.

Stockholm Paradigm is a theoretical framework that helps to predict the spread of parasites in response to changing conditions. The framework encompasses four driving forces: ecological fitting, the geographic mosaic theory of coevolution, taxon pulses, and the oscillation hypothesis.

symbiosis describes an evolved relationship between species. Symbiosis can be mutualistic, commensalistic, or parasitic.

tapirs are a group of odd-toed hoofed mammals with smooth skin and short, elephant-like trunks. There are five living species occurring in Central and South America and Southeast Asia.

taxonomic inventory is a list of species at a given site that characterizes and defines the ecological community.

taxonomy is the scientific process of describing, categorizing, and determining the evolutionary relationships of organisms. The commonly used categories are domain, phylum, class, order, family, genus, and species.

tectonic plates are pieces of the Earth's crust and mantle that move independently from each other and form continents and the ocean floor.

trematodes are flatworms characterized by having two suckers, one on the anterior end and the other in the middle of the body. All trematodes are parasites that use molluscs as their first intermediate host.

tuco-tucos are small subterranean rodents in the genus *Ctenomys* found in suitable habitats for burrowing in the southern portion of South America.

variable gene expression occurs when a gene produces an effect in an organism that changes depending on particular environmental conditions.

veliger is a free-swimming larval stage of a mollusc.

villi are small fingerlike projections that line the intestine of vertebrates.

Yanomami are Indigenous people living in the Amazon rainforests of Brazil and Venezuela.

Bibliography

Abollo, Elvira, Camino Gestal, Alfredo López, Ángel F. González, Angel Guerra, and Santiago Pascual. 1998. "Squid as Trophic Bridges for Parasite Flow within Marine Ecosystems: The Case of *Anisakis simplex* (Nematoda: Anisakidae), or When the Wrong Way Can Be Right." *African Journal of Marine Science* 20.

Agosta, Salvatore J., Niklas Janz, and Daniel R. Brooks. 2010. "How Specialists Can Be Generalists: Resolving the 'Parasite Paradox' and Implications for Emerging Infectious Disease." *Zoologia (Curitiba)* 27 (2): 151–62. https://doi.org/10.1590/S1984-46702010000200001.

Amin, Omar M. 2013. "Classification of the Acanthocephala." *Folia Parasitologica* 60 (4): 273–305. https://doi.org/10.14411/fp.2013.031.

Anderson, Roy Clayton, Alain Gabriel Chabaud, and Sheila Willmott, eds. 2009. *Keys to the Nematode Parasites of Vertebrates.* CABI.

Anderson, Sydney. 1997. "Mammals of Bolivia: Taxonomy and Distribution." *Bulletin of the American Museum of Natural History*, no. 231. http://digitallibrary.amnh.org/handle/2246/1620.

André, Amélie v Saint, Nathan M. Blackwell, Laurie R. Hall, Achim Hoerauf, Norbert W. Brattig, Lars Volkmann, Mark J. Taylor, et al. 2002. "The Role of Endosymbiotic Wolbachia Bacteria in the Pathogenesis of River Blindness." *Science* 295 (5561): 1892–95. https://doi.org/10.1126/science.1068732.

Anonymous. 1935. "*Ascaris* Infection and the Bore-Hole Latrine." *Indian Medical Gazette*, June.

Araújo, Adauto, Ana Maria Jansen, Karl Reinhard, and Luiz Fernando Ferreira. 2009. "Paleoparasitology of Chagas Disease: A Review." *Memórias Do Instituto Oswaldo Cruz* 104: 9–16. https://doi.org/10.1590/S0074-02762009000900004.

Araújo, Adauto, Adriana Rangel, and Luiz Fernando Rocha Ferreira. 1993. "Climatic Change in Northeastern Brazil: Paleoparasitological Data." *Memórias Do Instituto Oswaldo Cruz* 88 (4): 577–79.

Araújo, Adauto, Karl J. Reinhard, Luiz Fernando Ferreira, and Scott L. Gardner. 2008. "Parasites as Probes for Prehistoric Human Migrations?" *Trends in Parasitology* 24 (3): 112–15. https://doi.org/10.1016/j.pt.2007.11.007.

Araújo, Adauto, Karl J. Reinhard, Luiz Fernando Ferreira, Elisa Pucu, and Pedro Paulo Chieffi. 2013. "Paleoparasitology: The Origin of Human Parasites." *Arquivos de Neuro-Psiquiatria* 71 (9B): 722–26. https://doi.org/10.1590/0004-282X20130159.

Araujo, Sabrina B. L., Mariana Pires Braga, Daniel R. Brooks, Salvatore J. Agosta, Eric P. Hoberg, Francisco W. von Hartenthal, and Walter A. Boeger. 2015. "Understanding Host-Switching by Ecological Fitting." *PLOS ONE* 10 (10): e0139225. https://doi.org/10.1371/journal.pone.0139225.

Auld, Stuart K.J.R., and Matthew C. Tinsley. 2015. "The Evolutionary Ecology of Complex Life-cycle Parasites: Linking Phenomena with Mechanisms." *Heredity* 114 (2): 125–32. https://doi.org/10.1038/hdy.2014.84.

Baker, Robert J., and Robert D. Bradley. 2006. "Speciation in Mammals and the Genetic Species Concept." *Journal of Mammalogy* 87 (4): 643–62. https://doi.org/10.1644/06-MAMM-F-038R2.1.

Banyard, Ashley C., David Hayman, Nicholas Johnson, Lorraine McElhinney, and Anthony R. Fooks. 2011. "Chapter 12—Bats and Lyssaviruses." In *Advances in Virus Research Volume 79*, edited by Alan C. Jackson, 79:239–89. Research Advances in Rabies. Academic Press. https://doi.org/10.1016/B978-0-12-387040-7.00012-3.

Bar-On, Yinon M., Rob Phillips, and Ron Milo. 2018. "The Biomass Distribution on Earth." *Proceedings of the National Academy of Sciences* 115 (25): 6506–11.

Basáñez, María-Gloria, Sébastien D. S. Pion, Thomas S. Churcher, Lutz P. Breitling, Mark P. Little, and Michel Boussinesq. 2006. "River Blindness: A Success Story under Threat?" *PLOS Medicine* 3 (9): e371. https://doi.org/10.1371/journal.pmed.0030371.

Bataille, Arnaud, Iris I. Levin, and Eloisa H. R. Sari. 2018. "Colonization of Parasites and Vectors." In *Disease Ecology: Galapagos Birds and Their Parasites*, edited by Patricia G. Parker, 45–79. Social and Ecological Interactions in the Galapagos Islands. Cham: Springer International Publishing. https://doi.org/10.1007/978-3-319-65909-1_3.

Batsaikhan, Nyamsuren, Bayarbaatar Buuveibaatar, Bazaar Chimed, Oidov Enkhtuya, Davaa Galbrakh, Oyunsaikhan Ganbaatar, Badamjav Lkhagvasuren, et al. 2014. "Conserving the World's Finest Grassland amidst Ambitious National Development." *Conservation Biology* 28 (6): 1736–39. https://doi.org/10.1111/cobi.12297.

Bauer, Alexandre, Eleanor R. Haine, Marie-Jeanne Perrot-Minnot, and Thierry Rigaud. 2005. "The Acanthocephalan Parasite *Polymorphus minutus* Alters the Geotactic and Clinging Behaviours of Two Sympatric Amphipod Hosts: The Native *Gammarus pulex* and the Invasive *Gammarus roeseli*." *Journal of Zoology* 267 (1): 39–43. https://doi.org/10.1017/S0952836905007223.

Bavestrello, Giorgio, Attilio Arillo, Barbara Calcinai, Riccardo Cattaneo-Vietti, Carlo Cerrano, Elda Gaino, Antonella Penna, and Michele Sara. 2000. "Parasitic Diatoms inside Antarctic Sponges." *Biological Bulletin* 198 (1): 29–33.

Benedict, Russell, Patricia Freeman, Hugh Genoways, Freed B. Samson, and Fritz L. Knopf. 1996. "Prairie Legacies—Mammals." In *Prairie Conservation: Preserving North America's Most Endangered Ecosystem*, 149–66. Island Press.

Benton, Bruce, Jesse Bump, Azodoga Sékétéli, and Bernhard Liese. 2002. "Partnership and Promise: Evolution of the African River-Blindness Campaigns." *Annals of Tropical Medicine & Parasitology* 96 (sup1): S5–14. https://doi.org/10.1179/000349802125000619.

Bethel, William M., and John C. Holmes. 1973. "Altered Evasive Behavior and Responses to Light in Amphipods Harboring Acanthocephalan Cystacanths." *Journal of Parasitology* 59 (6): 945–56. https://doi.org/10.2307/3278623.

———. 1974. "Correlation of Development of Altered Evasive Behavior in *Gammarus lacustris* (Amphipoda) Harboring Cystacanths of *Polymorphus paradoxus* (Acanthocephala) with the Infectivity to the Definitive Host." *Journal of Parasitology* 60 (2): 272–74. https://doi.org/10.2307/3278463.

———. 1977. "Increased Vulnerability of Amphipods to Predation Owing to Altered Behavior Induced by Larval Acanthocephalans." *Canadian Journal of Zoology* 55 (1). https://doi.org/10.1139/z77-013.

Biggs, Alton. 2002. *Glencoe: Life Science*. New York: Glencoe/McGraw-Hill.

Black, Craig C. 1989. *Loss of Biological Diversity: A Global Crisis Requiring International Solutions. A Report to the National Science Board*. Vol. 89. Washington, D.C.: National Science Foundation.

Blakeslee, April M. H., Irit Altman, A. Whitman Miller, James E. Byers, Caitlin E. Hamer, and Gregory M. Ruiz. 2012. "Parasites and Invasions: A Biogeographic Examination of Parasites and Hosts in Native and Introduced Ranges." *Journal of Biogeography* 39 (3): 609–22. https://doi.org/10.1111/j.1365-2699.2011.02631.x.

Blanks, Jack, Frank Richards, F. Beltrán, R. Collins, Edmundo Álvarez, Guillermo Zea Flores, B. Bauler, et al. 1998. "The Onchocerciasis Elimination Program for the Americas: A History of Partnership." *Revista Panamericana de Salud Pública* 3 (6): 367–74. https://doi.org/10.1590/S1020-49891998000600002.

Bleed, Ann Salomon, and Charles Flowerday. 1989. *An Atlas of the Sand Hills*. Resource Atlas, No. 5. Conservation and Survey Division, Institute of Agriculture and Natural Resources, University of Nebraska-Lincoln.

Blend, Charles K., Norman O. Dronen, Gabor R. Racz, and Scott L. Gardner. 2017. "*Pseudopecoelus mccauleyi* n. sp. and *Podocotyle* sp. (Digenea: Opecoelidae) from the Deep Waters off Oregon and British Columbia with an Updated Key to the Species of *Pseudopecoelus* von Wicklen, 1946 and Checklist of Parasites from *Lycodes cortezianus* (Perciformes: Zoarcidae)." *Acta Parasitologica* 62 (2): 231–54. https://doi.org/10.1515/ap-2017-0031.

Boeger, Walter A., and Delane C. Kritsky. 1997. "Coevolution of the Monogenoidea (Platyhelminthes) Based on a Revised Hypothesis of Parasite Phylogeny." *International Journal for Parasitology* 27 (12): 1495–1511. https://doi.org/10.1016/S0020-7519(97)00140-9.

Bolek, Matthew G., and John Janovy Jr. 2008. "Alternative Life Cycle Strategies of *Megalodiscus temperatus* in Tadpoles and Metamorphosed Anurans." *Parasite* 15 (3): 396–401. https://doi.org/10.1051/parasite/2008153396.

Bolek, Matthew G., Andreas Schmidt-Rhaesa, L. Cristina De Villalobos, and Ben Hanelt. 2015. "Chapter 15—Phylum Nematomorpha." In *Thorp and Covich's Freshwater Invertebrates (Fourth Edition)*, edited by James H. Thorp and D. Christopher Rogers, 303–26. Boston: Academic Press. https://doi.org/10.1016/B978-0-12-385026-3.00015-2.

Bomberger, Mary L., Shelly L. Shields, A. Tyrone Harrison, and Kathleen H. Keeler. 1983. "Comparison of Old Field Succession on a Tallgrass Prairie and a Nebraska Sandhills Prairie." *Prairie Naturalist* 13 (1): 9–15.

Borup, Lance H., John S. Peters, and Christopher R. Sartori. 2003. "Onchocerciasis (River Blindness)." *Cutis* 72: 297–302.

Brant, Sara V., and Scott L. Gardner. 1997. "Two New Species of *Litomosoides* (Nemata: Onchocercidae) from *Ctenomys opimus* (Rodentia: Ctenomyidae) on the Altiplano of Bolivia." *Journal of Parasitology* 83 (4): 700–705. https://doi.org/10.2307/3284249.

———. 2000. "Phylogeny of Species of the Genus *Litomosoides* (Nematoda: Onchocercidae): Evidence of Rampant Host Switching." *Journal of Parasitology* 86 (3): 545–54. https://doi.org/10.1645/0022-3395(2000)086[0545:POSOTG]2.0.CO;2.

Brattig, Norbert W., Dietrich W. Büttner, and Achim Hoerauf. 2001. "Neutrophil Accumulation around Onchocerca Worms and Chemotaxis of Neutrophils Are Dependent on Wolbachia Endobacteria." *Microbes and Infection* 3 (6): 439–46. https://doi.org/10.1016/S1286-4579(01)01399-5.

Bray, Rodney A., David I. Gibson, and Arlene Jones, eds. 2008. *Keys to the Trematoda. Volume 3.* Wallingford: CABI.

Briones, Marcelo R. S., Ricardo P. Souto, Beatriz S. Stolf, and Bianca Zingales. 1999. "The Evolution of Two *Trypanosoma cruzi* Subgroups Inferred from rRNA Genes Can Be Correlated with the Interchange of American Mammalian Faunas in the Cenozoic and Has Implications to Pathogenicity and Host Specificity." *Molecular and Biochemical Parasitology* 104 (2): 219–32. https://doi.org/10.1016/S0166-6851(99)00155-3.

Brooks, Daniel R., and Walter A. Boeger. 2019. "Climate Change and Emerging Infectious Diseases: Evolutionary Complexity in Action." *Current Opinion in Systems Biology* 13 (February): 75–81. https://doi.org/10.1016/j.coisb.2018.11.001.

Brooks, Daniel R., and David R. Glen. 1982. "Pinworms and Primates: A Case Study in Coevolution." *Proceedings of the Helminthological Society of Washington* 49 (1): 76–85.

Brooks, Daniel R., and Eric P. Hoberg. 2000. "Triage for the Biosphere: The Need and Rationale for Taxonomic Inventories and Phylogenetic Studies of Parasites." *Comparative Parasitology* 67 (1): 1–25.

———. 2007. "How Will Global Climate Change Affect Parasite–Host Assemblages?" *Trends in Parasitology* 23 (12): 571–74. https://doi.org/10.1016/j.pt.2007.08.016.

Brooks, Daniel R., Eric P. Hoberg, and Walter A. Boeger. 2019. *The Stockholm Paradigm: Climate Change and Emerging Disease.* Chicago: University of Chicago Press.

Brooks, Daniel R., Eric P. Hoberg, Walter A. Boeger, Scott L. Gardner, Kurt E. Galbreath, Dávid Herczeg, Hugo H. Mejía-Madrid, S. Elizabeth Rácz, and Altangerel Tsogtsaikhan Dursahinhan. 2014. "Finding Them before They Find Us: Informatics, Parasites, and Environments in Accelerating Climate Change." *Comparative Parasitology* 81 (2): 155–64. https://doi.org/10.1654/4724b.1.

Brooks, Daniel R., and Deborah A. McLennan. 1993. *Parascript: Parasites and the Language of Evolution.* Smithsonian.

———. 2012. *The Nature of Diversity: An Evolutionary Voyage of Discovery.* University of Chicago Press.

Burge, W. E., and E. L. Burge. 1915. "The Protection of Parasites in the Digestive Tract against the Action of the Digestive Enzymes." *Journal of Parasitology* 1 (4): 179–83. https://doi.org/10.2307/3270806.

Bullard, Stephen A., and Robin M. Overstreet. 2008. "Digeneans as Enemies of Fishes." In *Fish Diseases Volume 2*, edited by Jorge C. Eiras, Helmut Segner, Thomas Wahli, and B. G. Kapoor, 817–976. Enfield, NH: Science Publishers.

Burnham, Gilbert. 1998. "Onchocerciasis." *Lancet* 351 (9112): 1341–46. https://doi.org/10.1016/S0140-6736(97)12450-3.

Cabrera-Gil, Susana, Abhay Deshmukh, Carlos Cervera-Estevan, Natalia Fraija-Fernández, Mercedes Fernández, and Francisco Javier Aznar. 2018. "Anisakis Infections in Lantern Fish (Myctophidae) from the Arabian Sea: A Dual Role for Lantern Fish in the Life Cycle of *Ani-*

sakis brevispiculata?" Deep Sea Research Part I: Oceanographic Research Papers 141 (November): 43–50. https://doi.org/10.1016/j.dsr.2018.08.004.

Caira, Janine N., and Kirsten Jensen, eds. 2017. *Planetary Biodiversity Inventory (2008–2017): Tapeworms from Vertebrate Bowels of the Earth*. Lawrence, KS: Natural History Museum, The University of Kansas.

Casiraghi, Maurizio, Odile Bain, Ricardo Guerrero, Coralie Martin, Vanessa Pocacqua, Scott L. Gardner, Alberto Franceschi, and Claudio Bandi. 2004. "Mapping the Presence of *Wolbachia pipientis* on the Phylogeny of Filarial Nematodes: Evidence for Symbiont Loss during Evolution." *International Journal for Parasitology* 34 (2): 191–203. https://doi.org/10.1016/j.ijpara.2003.10.004.

Calderon, Alfonso, Camilo Guzman, Jorge Salazar-Bravo, Luiz Figueiredo, Salim Mattar, and German Arrieta. 2016. "Viral Zoonoses That Fly with Bats: A Review." *MANTER: Journal of Parasite Biodiversity* 6 (September): 1–13.

Calisher, Charles H., James E. Childs, Hume E. Field, Kathryn V. Holmes, and Tony Schountz. 2006. "Bats: Important Reservoir Hosts of Emerging Viruses." *Clinical Microbiology Reviews* 19 (3): 531–45. https://doi.org/10.1128/CMR.00017-06.

Campbell, William C. 1981. "An Introduction to the Avermectins." *New Zealand Veterinary Journal* 29 (10): 174–78. https://doi.org/10.1080/00480169.1981.34836.

———. 2016. "Lessons from the History of Ivermectin and Other Antiparasitic Agents." *Annual Review of Animal Biosciences* 4 (1): 1–14. https://doi.org/10.1146/annurev-animal-021815-111209.

Campbell, William C., Richard W. Burg, Michael H. Fisher, and Richard A. Dybas. 1984. "The Discovery of Ivermectin and Other Avermectins." In *Pesticide Synthesis through Rational Approaches*, edited by Philip S. Magee, Gustave K. Kohn, and Julius J. Menn, 255: 5–20. ACS Symposium Series 255. Washington, D.C.: American Chemical Society. https://doi.org/10.1021/bk-1984-0255.ch001.

Carmichael, Lisa M., and Janice Moore. 1991. "A Comparison of Behavioral Alterations in the Brown Cockroach, *Periplaneta brunnea*, and the American Cockroach, *Periplaneta americana*, Infected with the Acanthocephalan, *Moniliformis moniliformis*." *Journal of Parasitology* 77 (6): 931–36. https://doi.org/10.2307/3282745.

Case, Ronald M., and Charles W. Ramsey. 1994. "Gophers, Pocket." In *Prevention and Control of Wildlife Damage*, edited by Scott E. Hygnstrom, Robert M. Timm, and Gary Eugene Larson, B-17. University of Nebraska Cooperative Extension, Institute of Agriculture and Natural Resources, University of Nebraska-Lincoln.

Ceballos, Gerardo, and Paul R. Ehrlich. 2002. "Mammal Population Losses and the Extinction Crisis." *Science* 296 (5569): 904–7. https://doi.org/10.1126/science.1069349.

Ceballos, Gerardo, Paul R. Ehrlich, Anthony D. Barnosky, Andrés García, Robert M. Pringle, and Todd M. Palmer. 2015. "Accelerated Modern Human–Induced Species Losses: Entering the Sixth Mass Extinction." *Science Advances* 1 (5): e1400253. https://doi.org/10.1126/sciadv.1400253.

Charles, Roxanne A., Sonia Kjos, Angela E. Ellis, John C. Barnes, and Michael J. Yabsley. 2012. "Southern Plains Woodrats (*Neotoma micropus*) from Southern Texas Are Important Reservoirs of Two Genotypes of *Trypanosoma cruzi* and Host of a Putative Novel *Trypanosoma*

Species." *Vector-Borne and Zoonotic Diseases* 13 (1): 22–30. https://doi.org/10.1089/vbz.2011 .0817.

Childs, James E., Thomas G. Ksiazek, Christina F. Spiropoulou, John W. Krebs, Sergey Morzunov, Gary O. Maupin, Kenneth L. Gage, et al. 1994. "Serologic and Genetic Identification of *Peromyscus maniculatus* as the Primary Rodent Reservoir for a New Hantavirus in the Southwestern United States." *Journal of Infectious Diseases* 169 (6): 1271–80. https://doi.org/10.1093 /infdis/169.6.1271.

Chimento, Nicolas, Federico Agnolin, and Agustin Martinelli. 2016. "Mesozoic Mammals from South America: Implications for Understanding Early Mammalian Faunas from Gondwana." In *Historia Evolutiva y Paleobiogeográfica de Los Vertebrados de América Del Sur*, edited by Frederico L. Agnolin, Gabriel L. Lio, Federico Brissón Egli, Nicolas R. Chimento, and Fernando E. Novas, 199–209. Contribuciones Del MACN 6. Buenos Aires, Brazil: Museo Argentino de Ciencias Naturales "Bernardino Rivadavia" e Instituto Nacional de Investigación de las Ciencias Naturales.

Chitsulo, Lester, Dirk Engels, Antonio Montresor, and Lorenzo Savioli. 2000. "The Global Status of Schistosomiasis and Its Control." *Acta Tropica* 77 (1): 41–51.

Choudhury, Anindo, M. Leopoldina Aguirre-Macedo, Stephen S. Curran, Margarita Ostrowski De Núñez, Robin M. Overstreet, Gerardo Pérez-Ponce de León, and Cláudia Portes Santos. 2016. "Trematode Diversity in Freshwater Fishes of the Globe II:'New World.'" *Systematic Parasitology* 93 (3): 271–82.

Clark, David B. 1979. "A Centipede Preying on a Nestling Rice Rat (*Oryzomys bauri*)." *Journal of Mammalogy* 60 (3): 654. https://doi.org/10.2307/1380119.

———. 1980. "Population Ecology of an Endemic Neotropical Island Rodent: *Oryzomys bauri* of Santa Fe Island, Galapagos, Ecuador." *Journal of Animal Ecology* 49 (1): 185–98. https://doi .org/10.2307/4283.

Clark, Deborah A., and David B. Clark. 1981. "Effects of Seed Dispersal by Animals on the Regeneration of *Bursera graveolens* (Burseraceae) on Santa Fe Island, Galápagos." *Oecologia* 49 (1): 73–75. https://doi.org/10.1007/BF00376900.

Clark, J. Desmond, and Hiro Kurashina. 1979. "Hominid Occupation of the East-Central Highlands of Ethiopia in the Plio–Pleistocene." *Nature* 282 (5734): 33–39. https://doi.org/10 .1038/282033a0.

Clark, Nicola, and Simon Wallis. 2017. "Flamingos, Salt Lakes and Volcanoes: Hunting for Evidence of Past Climate Change on the High Altiplano of Bolivia." *Geology Today* 33 (3): 101–7. https://doi.org/10.1111/gto.12186.

Coca-Salazar, Alejandro, Huber Villca, Mauricio Torrico, and Fernando D. Alfaro. 2016. "Plant Communities on the Islands of Two Altiplanic Salt Lakes in the Andean Region of Bolivia." *Check List* 12 (5): 1975. https://doi.org/10.15560/12.5.1975.

Conn, Jan E., Richard C. Wilkerson, M. Nazaré O. Segura, Raimundo T. L. de Souza, Carl D. Schlichting, Robert A. Wirtz, and Marinete M. Póvoa. 2002. "Emergence of a New Neotropical Malaria Vector Facilitated by Human Migration and Changes in Land Use." *American Journal of Tropical Medicine and Hygiene* 66 (1): 18–22. https://doi.org/10.4269/ajtmh.2002.66.18.

Cook, Joseph A., Kurt E. Galbreath, Kayce C. Bell, Mariel L. Campbell, Suzanne Carrière, Jocelyn P. Colella, Natalie G. Dawson, et al. 2016. "The Beringian Coevolution Project: Holistic

Collections of Mammals and Associated Parasites Reveal Novel Perspectives on Evolutionary and Environmental Change in the North." *Arctic Science* 3: 585–617. https://doi.org/10.1139/as-2016-0042.

Cook, Joseph A., Eric P. Hoberg, Anson Koehler, Heikki Henttonen, Lotta Wickström, Voitto Haukisalmi, Kurt Galbreath, et al. 2005. "Beringia: Intercontinental Exchange and Diversification of High Latitude Mammals and Their Parasites during the Pliocene and Quaternary." *Mammal Study* 30 (Supplement): S33–44. https://doi.org/10.3106/1348-6160(2005)30[33:BIEADO]2.0.CO;2.

COSEWIC. 2005. "COSEWIC Assessment and Update Status Report on the Fin Whale *Balaenoptera physalus* in Canada." Committee on the Status of Endangered Wildlife in Canada, Ottawa, ON. www.sararegistry.gc.ca/status/status_e.cfm.

Cotton, James A., Sasisekhar Bennuru, Alexandra Grote, Bhavana Harsha, Alan Tracey, Robin Beech, Stephen R. Doyle, et al. 2016. "The Genome of *Onchocerca volvulus*, Agent of River Blindness." *Nature Microbiology* 2 (2): 1–12. https://doi.org/10.1038/nmicrobiol.2016.216.

Cox, Frank E. G. 2002. "History of Human Parasitology." *Clinical Microbiology Reviews* 15 (4): 595–612. https://doi.org/10.1128/CMR.15.4.595-612.2002.

Craig, Philip. 2003. "*Echinococcus multilocularis*." *Current Opinion in Infectious Diseases* 16 (5): 437–44.

Crompton, David William Thomasson, and Brent B. Nickol. 1985. *Biology of the Acanthocephala*. Cambridge University Press.

Dailey, Murray D., Frances M. D. Gulland, Linda J. Lowenstine, Paul Silvagni, and Daniel Howard. 2000. "Prey, Parasites and Pathology Associated with the Mortality of a Juvenile Gray Whale (*Eschrichtius robustus*) Stranded along the Northern California Coast." *Diseases of Aquatic Organisms* 42 (2): 111–17. https://doi.org/10.3354/dao042111.

Dailey, Murray, and Wolfgang Vogelbein. 1991. "Parasite Fauna of 3 Species of Antarctic Whales with Reference to Their Use as Potential Stock Indicators." *Fishery Bulletin* 89 (3): 355–65.

Dawkins, Richard. 1982. *The Extended Phenotype: The Gene as the Unit of Selection*. Oxford: Oxford University Press.

De Baets, Kenneth, Paula Dentzien-Dias, Ieva Upeniece, Olivier Verneau, and Philip C. J. Donoghue. 2015. "Constraining the Deep Origin of Parasitic Flatworms and Host-Interactions with Fossil Evidence." In *Advances in Parasitology* 90: 93–135. Academic Press. https://doi.org/10.1016/bs.apar.2015.06.002.

Després, Laurence, Daniéle Imbert-Establet, and Monique Monnerot. 1993. "Molecular Characterization of Mitochondrial DNA Provides Evidence for the Recent Introduction of *Schistosoma mansoni* into America." *Molecular and Biochemical Parasitology* 60 (2): 221–29. https://doi.org/10.1016/0166-6851(93)90133-I.

Detwiler, Jillian, and John Janovy Jr. 2008. "The Role of Phylogeny and Ecology in Experimental Host Specificity: Insights from a Eugregarine–Host System." *Journal of Parasitology* 94 (1): 7–12. https://doi.org/10.1645/GE-1308.1.

Di Bella, Stefano, Niccolò Riccardi, Daniele Roberto Giacobbe, and Roberto Luzzati. 2018. "History of Schistosomiasis (Bilharziasis) in Humans: From Egyptian Medical Papyri to Molecular Biology on Mummies." *Pathogens and Global Health* 112 (5): 268–73. https://doi.org/10.1080/20477724.2018.1495357.

Dobson, Andy P. 1988. "Restoring Island Ecosystems: The Potential of Parasites to Control Introduced Mammals." *Conservation Biology* 2 (1): 31–39. https://doi.org/10.1111/j.1523-1739.1988.tb00333.x.

Dounias, Edmond, and Takanori Oishi. 2016. "Inland Traditional Capture Fisheries in the Congo Basin: Introduction." *Revue d'Ethnoécologie* 10: 1–7. https://doi.org/10.4000/ethnoecologie.2882.

Dowler, Robert C., Darin S. Carroll, and Cody W. Edwards. 2000. "Rediscovery of Rodents (Genus *Nesoryzomys*) Considered Extinct in the Galápagos Islands." *Oryx* 34 (2): 109–17. https://doi.org/10.1046/j.1365-3008.2000.00104.x.

Ducatez, Simon, Louis Lefebvre, Ferran Sayol, Jean-Nicolas Audet, and Daniel Sol. 2020. "Host Cognition and Parasitism in Birds: A Review of the Main Mechanisms." *Frontiers in Ecology and Evolution* 8 (102). https://doi.org/10.3389/fevo.2020.00102.

Duclos, Laura M., Bradford J. Danner, and Brent B. Nickol. 2006. "Virulence of *Corynosoma constrictum* (Acanthocephala: Polymorphidae) in *Hyalella azteca* (Amphipoda) throughout Parasite Ontogeny." *Journal of Parasitology* 92 (4): 749–55. https://doi.org/10.1645/GE-770R.1.

Dunn, Frederick L. 1963. "Acanthocephalans and Cestodes of South American Monkeys and Marmosets." *Journal of Parasitology* 49 (5): 717–22. https://doi.org/10.2307/3275912.

Dunne, Jennifer A., Kevin D. Lafferty, Andrew P. Dobson, Ryan F. Hechinger, Armand M. Kuris, Neo D. Martinez, John P. McLaughlin, Kim N. Mouritsen, Robert Poulin, and Karsten Reise. 2013. "Parasites Affect Food Web Structure Primarily through Increased Diversity and Complexity." *PLOS Biology* 11 (6): e1001579. https://doi.org/10.1371/journal.pbio.1001579.

Dunnum, Jonathan L., Richard Yanagihara, Karl M. Johnson, Blas Armien, Nyamsuren Batsaikhan, Laura Morgan, and Joseph A. Cook. 2017. "Biospecimen Repositories and Integrated Databases as Critical Infrastructure for Pathogen Discovery and Pathobiology Research." *PLOS Neglected Tropical Diseases* 11 (1): e0005133. https://doi.org/10.1371/journal.pntd.0005133.

Ďuriš, Zdeněk, Ivona Horká, Petr Jan Juračka, Adam Petrusek, and Floyd Sandford. 2011. "These Squatters Are Not Innocent: The Evidence of Parasitism in Sponge-Inhabiting Shrimps." *PLOS ONE* 6 (7): e21987. https://doi.org/10.1371/journal.pone.0021987.

Dursahinhan, Altangerel Tsogtsaikhan, Batsaikhan Nyamsuren, Danielle Marie Tufts, and Scott Lyell Gardner. 2017. "A New Species of *Catenotaenia* (Cestoda: Catenotaeniidae) from *Pygeretmus pumilio* Kerr, 1792 from the Gobi of Mongolia." *Comparative Parasitology* 84 (2): 124–34. https://doi.org/10.1654/1525-2647-84.2.124.

Duszynski, Donald W., Matthew G. Bolek, and Steve J. Upton. 2007. *Coccidia (Apicomplexa: Eimeriidae) of the Amphibians of the World*. Magnolia Press.

Duszynski, Donald W., Jana Kvičerová, and R. Scott Seville. 2018. *The Biology and Identification of the Coccidia (Apicomplexa) of Carnivores of the World*. Academic Press.

Egoscue, Harold J. 1960. "Laboratory and Field Studies of the Northern Grasshopper Mouse." *Journal of Mammalogy* 41 (1): 99–110. https://doi.org/10.2307/1376521.

Ehrlich, Paul R., and Peter H. Raven. 1964. "Butterflies and Plants: A Study in Coevolution." *Evolution* 18 (4): 586–608. https://doi.org/10.2307/2406212.

Elman, Cheryl, Robert A McGuire, and Barbara Wittman. 2014. "Extending Public Health: The Rockefeller Sanitary Commission and Hookworm in the American South." *American Journal of Public Health* 104 (1): 47–58.

Esch, Gerald W. 2004. *Parasites, People, and Places: Essays on Field Parasitology.* Cambridge University Press.

———. 2007. *Parasites and Infectious Disease: Discovery by Serendipity and Otherwise.* Cambridge University Press.

Faith, J. Tyler, John Rowan, and Andrew Du. 2019. "Early Hominins Evolved within Non-Analog Ecosystems." *Proceedings of the National Academy of Sciences* 116 (43): 21478–83.

Fay, Francis H. 1973. "The Ecology of *Echinococcus multilocularis* Leuckart, 1863, (Cestoda: Taeniidae) on St. Lawrence Island, Alaska—I. Background and Rationale." *Annales de Parasitologie Humaine et Comparée* 48 (4): 523–42. https://doi.org/10.1051/parasite/1973484523.

Ferreira, Luiz Fernando, Adauto Araújo, Ulisses Confalonieri, Marcia Chame, and Delir Corrêa Gomes. 1991. "*Trichuris* Eggs in Animal Coprolites Dated from 30,000 Years Ago." *Journal of Parasitology* 77 (3): 491–93.

Ferreira, Luiz Fernando, Karl J. Reinhard, and Adauto Araújo, eds. 2014. *Foundations of Paleoparasitology.* Rio de Janeiro: Editora FIOCRUZ.

Flynn, John J., André R. Wyss, and Reynaldo Charrier. 2007. "South America's Missing Mammals." *Scientific American* 296 (5): 68–75.

Freeman, Patricia W. 1998. "Mammals." In *An Atlas of the Sand Hills*, 3rd ed., edited by Ann Salomon Bleed and Charles Flowerday, 193–200. Resource Atlas, no. 5b. Lincoln, Nebraska: Conservation and Survey Division, Institute of Agriculture and Natural Resources, University of Nebraska-Lincoln.

Frias, Liesbeth, Hideo Hasegawa, Danica J. Stark, Milena Salgado Lynn, Senthilvel K.S.S. Nathan, Tock H. Chua, Benoit Goossens, Munehiro Okamoto, and Andrew J. J. MacIntosh. 2019. "A Pinworm's Tale: The Evolutionary History of *Lemuricola (Protenterobius) nycticebi*." *International Journal for Parasitology: Parasites and Wildlife* 8 (April): 25–32. https://doi.org/10.1016/j.ijppaw.2018.11.009.

Friend, Milton, J. Christian Franson, and Elizabeth A. Ciganovich, eds. 1999. *Field Manual of Wildlife Diseases: General Field Procedures and Diseases of Birds.* Washington, D.C.: US Geological Survey.

Frey, Jennifer K., Terry L. Yates, Donald W. Duszynski, William L. Gannon, and Scott L. Gardner. 1992. "Designation and Curatorial Management of Type Host Specimens (Symbiotypes) for New Parasite Species." *Journal of Parasitology* 78 (5): 930–32.

Gabet, Emmanuel J. 2000. "Gopher Bioturbation: Field Evidence for Non-Linear Hillslope Diffusion." *Earth Surface Processes and Landforms* 25 (13): 1419–28. https://doi.org/10.1002/1096-9837(200012)25:13<1419::AID-ESP148>3.0.CO;2-1.

Galaktionov, Kirill V., and Andrej A. Dobrovolskij. 2003. "Organization of Parthenogenetic and Hermaphroditic Generations of Trematodes." In *The Biology and Evolution of Trematodes*, by Kirill V. Galaktionov and Andrej A. Dobrovolskij, 1–213. Dordrecht, The Netherlands: Kluwer Academic.

Galbreath, Kurt E., Eric P. Hoberg, Joseph A. Cook, Blas Armién, Kayce C. Bell, Mariel L. Campbell, Jonathan L. Dunnum, Altangerel T. Dursahinhan, Ralph P. Eckerlin, and Scott L. Gardner. 2019. "Building an Integrated Infrastructure for Exploring Biodiversity: Field Collections and Archives of Mammals and Parasites." *Journal of Mammalogy* 100 (2): 382–93.

Gamboa, María I., Graciela T. Navone, Alicia B. Orden, María F. Torres, Luis E. Castro, and Evelia E. Oyhenart. 2011. "Socio-Environmental Conditions, Intestinal Parasitic Infections and Nutritional Status in Children from a Suburban Neighborhood of La Plata, Argentina." *Acta Tropica* 118 (3): 184–89. https://doi.org/10.1016/j.actatropica.2009.06.015.

Ganzorig, Sumiya, Nyamsuren Batsaikhan, Yuzaburo Oku, and Masao Kamiya. 2002. "A New Nematode, *Soboliphyme ataahai* sp. n. (Nematoda: Soboliphymidae) from Laxmann's Shrew, *Sorex caecutiens* Laxmann, 1788 in Mongolia." *Parasitology Research* 89 (1): 44–48. https://doi.org/10.1007/s00436-002-0725-1.

Ganzorig, Sumiya, Nyamsuren Batsaikhan, Ravchig Samiya, Yasuyuki Morishima, Yuzaburo Oku, and Masao Kamiya. 1999. "A Second Record of Adult *Ascarops strongylina* (Rudolphi, 1819) (Nematoda: Spirocercidae) in a Rodent Host." *Journal of Parasitology* 85 (2): 283–85. https://doi.org/10.2307/3285633.

Ganzorig, Sumiya, Damdin Sumiya, Nyamsuren Batsaikhan, Rolf Schuster, Yuzaburo Oku, and Masao Kamiya. 1998. "New Findings of Metacestodes and a Pentastomid from Rodents in Mongolia." *Journal of the Helminthological Society of Washington* 65 (1): 74–81.

Gardner, Scott L. 1991. "Phyletic Coevolution between Subterranean Rodents of the Genus *Ctenomys* (Rodentia: Hystricognathi) and Nematodes of the Genus *Paraspidodera* (Heterakoidea: Aspidoderidae) in the Neotropics: Temporal and Evolutionary Implications." *Zoological Journal of the Linnean Society* 102 (2): 169–201. https://doi.org/10.1111/j.1096-3642.1991.tb00288.x.

———. 2001. "Worms, Nematoda." In *Encyclopedia of Biodiversity, Volume 5*, edited by Simon A. Levin, 843–62. San Diego: Academic Press.

Gardner, Scott L., and Mariel L. Campbell. 1992a. "A New Species of *Linstowia* (Cestoda: Anoplocephalidae) from Marsupials in Bolivia." *Journal of Parasitology* 78 (5): 795–99. https://doi.org/10.2307/3283306.

———. 1992b. "Parasites as Probes for Biodiversity." *Journal of Parasitology* 78 (4): 596–600. https://doi.org/10.2307/3283534.

Gardner, Scott L., Altangerel T. Dursahinhan, Mariel L. Campbell, and S. Elizabeth Rácz. 2020. "A New Genus and Two New Species of Unarmed Hymenolepidid Cestodes (Cestoda: Hymenolepididae) from Geomyid Rodents in Mexico and Costa Rica." *Zootaxa* 4766 (2): 358–76. https://doi.org/10.11646/zootaxa.4766.2.5.

Gardner, Scott L., and Donald W. Duszynski. 1990. "Polymorphism of Eimerian Oocysts Can Be a Problem in Naturally Infected Hosts: An Example from Subterranean Rodents in Bolivia." *Journal of Parasitology* 76 (6): 805–11.

Gardner, Scott L., and Jean-Pierre Hugot. 1995. "A New Pinworm, *Didelphoxyuris thylamisis* n. gen., n. sp. (Nematoda: Oxyurida) from *Thylamys elegans* (Waterhouse, 1839) (Marsupialia: Didelphidae) in Bolivia." *Research and Reviews in Parasitology* 55 (4): 139–47.

Gardner, Scott L., Brent A. Luedders, and Donald W. Duszynski. 2014. "*Hymenolepis robertrauschi* n. sp. from Grasshopper Mice *Onychomys* spp. in New Mexico and Nebraska, U.S.A." *Occasional Papers, Museum of Texas Tech University*, no. 322 (March): 1–10.

Gardner, Scott L., Robert L. Rausch, and Otto Carlos Jordan Camacho. 1988. "*Echinococcus vogeli* Rausch and Bernstein, 1972, from the Paca, *Cuniculus paca* L. (Rodentia: Dasyproctidae), in the Departamento de Santa Cruz, Bolivia." *Journal of Parasitology* 74 (3): 399–402. https://doi.org/10.2307/3282045.

Gardner, Scott L., Altangerel Dursahinhan, Gábor Rácz, Nyamsuren Batsaikhan, Sumiya Ganzorig, David Tinnin, Darmaa Damdinbazar, et al. 2013. "Sylvatic Species of *Echinococcus* from Rodent Intermediate Hosts in Asia and South America." *Occasional Papers, Museum of Texas Tech University* 318 (October): 1–13.

Gardner, Scott L., Jorge Salazar-Bravo, and Joseph A. Cook. 2014. "New Species of *Ctenomys* Blainville 1826 (Rodentia: Ctenomyidae) from the Lowlands and Central Valleys of Bolivia." *Special Publications–Museum of Texas Tech University* 62: 1–34.

Gardner, Scott L., Nathan A. Seggerman, Nyamsuren Batsaikhan, Sumiya Ganzorig, David S. Tinnin, and Donald W. Duszynski. 2009. "Coccidia (Apicomplexa: Eimeriidae) from the Lagomorph *Lepus tolai* in Mongolia." *Journal of Parasitology* 95 (6): 1451–54. https://doi.org/10.1645/GE-2137.1.

Gardner, Scott L., and Peter T. Thew. 2006. "Redescription of *Cryptocotyle thapari* McIntosh, 1953 (Trematoda: Heterophyidae), in the River Otter *Lutra longicaudis* from Bolivia." *Comparative Parasitology* 73 (1): 20–23. https://doi.org/10.1654/0001.1.

Gardner, Scott L., Steve Upton, C. R. Lambert, and O. C. Jordan. 1991. "Redescription of *Eimeria escomeli* (Rastegaieff, 1930) from *Myrmecophaga tridactyla*, and a First Report from Bolivia." *Journal of the Helminthological Society of Washington* 58 (1): 16–18.

Garey, James R., Andreas Schmidt-Rhaesa, Thomas J. Near, and Steven A. Nadler. 1998. "The Evolutionary Relationships of Rotifers and Acanthocephalans." In *Rotifera VIII: A Comparative Approach*, edited by Elizabeth S. Wurdak, Robert L. Wallace, and Hendrik Segers, 83–91. Springer.

Gemmell, Michael Alexander. 1959. "The Fox as a Definitive Host of *Echinococcus* and Its Role in the Spread of Hydatid Disease." *Bulletin of the World Health Organization* 20 (1): 87–99.

Genoways, Hugh H., Meredith J. Hamilton, Darin M. Bell, Ryan R. Chambers, and Robert D. Bradley. 2008. "Hybrid Zones, Genetic Isolation, and Systematics of Pocket Gophers (Genus *Geomys*) in Nebraska." *Journal of Mammalogy* 89 (4): 826–36. https://doi.org/10.1644/07-MAMM-A-408.1.

Georgiev, Boyko B., Rodney A. Bray, D. Timothy, and J. Littlewood. 2006. "Cestodes of Small Mammals: Taxonomy and Life Cycles." In *Micromammals and Macroparasites: From Evolutionary Ecology to Management*, edited by Serge Morand, Boris R. Krasnov, and Robert Poulin, 29–62. Tokyo: Springer Japan. https://doi.org/10.1007/978-4-431-36025-4_3.

Gibson, David I., Arlene Jones, and Rodney A. Bray, eds. 2002. *Keys to the Trematoda. Volume 1.* CABI.

Goater, Timothy M., Cameron P. Goater, and Gerald W. Esch. 2014. *Parasitism: The Diversity and Ecology of Animal Parasites.* 2nd ed. Cambridge University Press.

Goble, R. J., Joseph A. Mason, David B. Loope, and James B. Swinehart. 2004. "Optical and Radiocarbon Ages of Stacked Paleosols and Dune Sands in the Nebraska Sand Hills, USA." *Quaternary Science Reviews* 23 (9): 1173–82. https://doi.org/10.1016/j.quascirev.2003.09.009.

Gonçalves, Marcelo Luiz Carvalho, Adauto Araújo, and Luiz Fernando Ferreira. 2003. "Human Intestinal Parasites in the Past: New Findings and a Review." *Memórias Do Instituto Oswaldo Cruz* 98 (Suppl. 1): 103–18. https://doi.org/10.1590/S0074-02762003000900016.

Gonzalez-Astudillo, Viviana, Héctor Ramírez-Chaves, Joerg Henning, and Thomas Gillespie. 2016. "Current Knowledge of Studies of Pathogens in Colombian Mammals." *MANTER: Journal of Parasite Biodiversity* 4 (September): 1–13.

Gosselin, David C., Steve Sibray, and Jerry Ayers. 1994. "Geochemistry of K-Rich Alkaline Lakes, Western Sandhills, Nebraska, USA." *Geochimica et Cosmochimica Acta* 58 (5): 1403–18.

Gotelli, Nicholas J., and Janice Moore. 1992. "Altered Host Behaviour in a Cockroach-Acanthocephalan Association." *Animal Behaviour* 43 (6): 949–59. https://doi.org/10.1016/S0003-3472(06)80008-4.

Gubanov, N. M. 1951. "A Giant Nematode from the Placenta of Cetaceans *Placentonema gigantissima* nov. gen., nov. sp." *Doklady Akademii Nauk SSSR* 77 (6): 1123–25.

Guerrero, Ricardo, Coralie Martin, Scott L. Gardner, and Odile Bain. 2002. "New and Known Species of *Litomosoides* (Nematoda: Filarioidea): Important Adult and Larval Characters and Taxonomic Changes." *Comparative Parasitology* 69 (2): 177–95. https://doi.org/10.1654/1525-2647(2002)069[0177:NAKSOL]2.0.CO;2.

Guhl, Felipe, Arthur Auderheide, and Juan David Ramírez. 2014. "From Ancient to Contemporary Molecular Eco-Epidemiology of Chagas Disease in the Americas." *International Journal for Parasitology*, ICOPA XIII, 44 (9): 605–12. https://doi.org/10.1016/j.ijpara.2014.02.005.

Gustafsson, Margaretha K. S., Krister Eriksson, and Annika Hydén. 1995. "Never Ending Growth and a Growth Factor. II. Immunocytochemical Evidence for the Presence of Epidermal Growth Factor in a Tapeworm." *Hydrobiologia* 305 (1): 229–33. https://doi.org/10.1007/BF00036394.

Gustavsen, Ken, Adrian Hopkins, and Mauricio Sauerbrey. 2011. "Onchocerciasis in the Americas: From Arrival to (near) Elimination." *Parasites & Vectors* 4 (1): 205. https://doi.org/10.1186/1756-3305-4-205.

Halton, David W., and Margaretha K. S. Gustafsson. 1996. "Functional Morphology of the Platyhelminth Nervous System." *Parasitology* 113 (S1): S47–72. https://doi.org/10.1017/S0031182000077891.

Hamilton, Patrick B., Marta M. G. Teixeira, and Jamie R. Stevens. 2012. "The Evolution of *Trypanosoma cruzi*: The 'Bat Seeding' Hypothesis." *Trends in Parasitology* 28 (4): 136–41. https://doi.org/10.1016/j.pt.2012.01.006.

Hansson, Lennart, and Heikki Henttonen. 1988. "Rodent Dynamics as Community Processes." *Trends in Ecology & Evolution* 3 (8): 195–200. https://doi.org/10.1016/0169-5347(88)90006-7.

Harris, Donna B., Stephen D. Gregory, and David W. Macdonald. 2006. "Space Invaders? A Search for Patterns Underlying the Coexistence of Alien Black Rats and Galápagos Rice Rats." *Oecologia* 149 (2): 276. https://doi.org/10.1007/s00442-006-0447-7.

Hasegawa, Hideo. 1999. "Phylogeny, Host-Parasite Relationship and Zoogeography." *Korean Journal of Parasitology* 37 (4): 197–213. https://doi.org/10.3347/kjp.1999.37.4.197.

Hechinger, Ryan F., Kate L. Sheehan, and Andrew V. Turner. 2019. "Metabolic Theory of Ecology Successfully Predicts Distinct Scaling of Ectoparasite Load on Hosts." *Proceedings of the Royal Society B: Biological Sciences* 286 (1917): 20191777. https://doi.org/10.1098/rspb.2019.1777.

Hermosilla, Carlos, Liliana M. R. Silva, Rui Prieto, Sonja Kleinertz, Anja Taubert, and Monica A. Silva. 2015. "Endo- and Ectoparasites of Large Whales (Cetartiodactyla: Balaenopteridae, Physeteridae): Overcoming Difficulties in Obtaining Appropriate Samples by Non- and Minimally-Invasive Methods." *International Journal for Parasitology: Parasites and Wildlife* 4 (3): 414–20. https://doi.org/10.1016/j.ijppaw.2015.11.002.

Hindsbo, Ole. 1972. "Effects of *Polymorphus* (Acanthocephala) on Colour and Behaviour of *Gammarus lacustris*." *Nature* 238 (5363): 333.

Hoberg, Eric P. 1986. "Evolution and Historical Biogeography of a Parasite–Host Assemblage: *Alcataenia* spp. (Cyclophyllidea: Dilepididae) in Alcidae (Charadriiformes)." *Canadian Journal of Zoology* 64 (11): 2576–89.

———. 1992. "Congruent and Synchronic Patterns in Biogeography and Speciation among Seabirds, Pinnipeds, and Cestodes." *Journal of Parasitology* 78 (4): 601–15.

———. 1997. "Phylogeny and Historical Reconstruction: Host–Parasite Systems as Keystones in Biogeography and Ecology." In *Biodiversity II: Understanding and Protecting Our Biological Resources*, edited by Marjorie L. Reaka-Kudla, Don E. Wilson, and Edward O. Wilson, 243–61. Washington, D.C.: Joseph Henry Press.

———. 1999. "Phylogenetic Analysis among the Families of the Cyclophyllidea (Eucestoda) Based on Comparative Morphology, with New Hypotheses for Co-Evolution in Vertebrates." *Systematic Parasitology* 42 (1): 51–73. https://doi.org/10.1023/A:1006100629059.

———. 2002a. "Foundations for an Integrative Parasitology: Collections, Archives, and Biodiversity Informatics." *Comparative Parasitology* 69 (2): 124–31. https://doi.org/10.1654/1525 -2647(2002)069[0124:FFAIPC]2.0.CO;2.

———. 2002b. "*Taenia* Tapeworms: Their Biology, Evolution and Socioeconomic Significance." *Microbes and Infection* 4 (8): 859–66. https://doi.org/10.1016/S1286-4579(02)01606-4.

———. 2006. "Phylogeny of *Taenia*: Species Definitions and Origins of Human Parasites." *Parasitology International* 55 (January): S23–30. https://doi.org/10.1016/j.parint.2005.11.049.

———. 2014. "Robert Lloyd Rausch—A Life in Nature and Field Biology: 1921–2012." *Journal of Parasitology* 100 (4): 547–52.

Hoberg, Eric P., and A. Adams. 2000. "Phylogeny, History and Biodiversity: Understanding Faunal Structure and Biogeography in the Marine Realm." *Bulletin of the Scandinavian Society for Parasitology* 10 (2): 19–37.

Hoberg, Eric P., Salvatore J. Agosta, Walter A. Boeger, and Daniel R. Brooks. 2015. "An Integrated Parasitology: Revealing the Elephant through Tradition and Invention." *Trends in Parasitology* 31 (4): 128–33. https://doi.org/10.1016/j.pt.2014.11.005.

Hoberg, Eric P., Nancy L. Alkire, Alen de Queiroz, and Arlene Jones. 2001. "Out of Africa: Origins of the *Taenia* Tapeworms in Humans." *Proceedings of the Royal Society of London. Series B: Biological Sciences* 268 (1469): 781–87. https://doi.org/10.1098/rspb.2000.1579.

Hoberg, Eric P., and Daniel R. Brooks. 2008. "A Macroevolutionary Mosaic: Episodic Host-Switching, Geographical Colonization and Diversification in Complex Host–Parasite Systems." *Journal of Biogeography* 35 (9): 1533–50. https://doi.org/10.1111/j.1365-2699.2008 .01951.x.

———. 2010. "Beyond Vicariance: Integrating Taxon Pulses, Ecological Fitting, and Oscillation in Evolution and Historical Biogeography." In *The Biogeography of Host-Parasite Interactions*, edited by Serge Morand and Boris R. Krasnov, 7–20. Oxford: Oxford University Press.

———. 2015. "Evolution in Action: Climate Change, Biodiversity Dynamics and Emerging Infectious Disease." *Philosophical Transactions of the Royal Society B: Biological Sciences* 370 (1665): 20130553. https://doi.org/10.1098/rstb.2013.0553.

Hoberg, Eric P., Daniel R Brooks, and Douglas Siegel-Causey. 1997. "Host-Parasite Co-Speciation: History, Principles, and Prospects." In *Host–Parasite Evolution: General Principles and Avian Models*, edited by Dale H. Clayton and Janice Moore, 212–35. Oxford University Press.

Hoberg, Eric P., Joseph A. Cook, Salvatore J. Agosta, Walter A. Boeger, Kurt E. Galbreath, Sauli Laaksonen, Susan J. Kutz, and Daniel R. Brooks. 2017. "Arctic Systems in the Quaternary: Ecological Collision, Faunal Mosaics and the Consequences of a Wobbling Climate." *Journal of Helminthology* 91 (4): 409–21. https://doi.org/10.1017/S0022149X17000347.

Hoberg, Eric P., Arlene Jones, Robert L. Rausch, Keeseon S. Eom, and Scott L. Gardner. 2000. "A Phylogenetic Hypothesis for Species of the Genus *Taenia* (Eucestoda: Taeniidae)." *Journal of Parasitology* 86 (1): 89–98.

Hoberg, Eric P., Susan J. Kutz, Kurt E. Galbreath, and Joseph A. Cook. 2003. "Arctic Biodiversity: From Discovery to Faunal Baselines—Revealing the History of a Dynamic Ecosystem." *Journal of Parasitology* 89 (Suppl): S84–95.

Hoberg, Eric P., Jean Mariaux, Jean-Lou Justine, Daniel R. Brooks, and Peter J. Weekes. 1997. "Phylogeny of the Orders of the Eucestoda (Cercomeromorphae) Based on Comparative Morphology: Historical Perspectives and a New Working Hypothesis." *Journal of Parasitology* 83 (6): 1128–47.

Hoberg, Eric P., Kirsten J. Monsen, Susan J. Kutz, and Michael S. Blouin. 1999. "Structure, Biodiversity, and Historical Biogeography of Nematode Faunas in Holarctic Ruminants: Morphological and Molecular Diagnoses for *Teladorsagia boreoarcticus* n. sp. (Nematoda: Ostertagiinae), a Dimorphic Cryptic Species in Muskoxen (*Ovibos moschatus*)." *Journal of Parasitology* 85 (5): 910–34.

Hoberg, Eric P., Patricia A. Pilitt, and Kurt E. Galbreath. 2009. "Why Museums Matter: A Tale of Pinworms (Oxyuroidea: Heteroxynematidae) among Pikas (*Ochotona princeps* and *O. collaris*) in the American West." *Journal of Parasitology* 95 (2): 490–501. https://doi.org/10.1645/GE-1823.1.

Hoberg, Eric P., Lydden Polley, Emily J. Jenkins, Susan J. Kutz, Alasdair M. Veitch, and Brett T. Elkin. 2008. "Integrated Approaches and Empirical Models for Investigation of Parasitic Diseases in Northern Wildlife." *Emerging Infectious Diseases* 14 (1): 10–17. https://doi.org/10.3201/eid1401.071119.

Hoeppli, R., and I. H. Ch'iang. 1940. "Selections from Old Chinese Medical Literature on Various Subjects of Helminthological Interest." *Chinese Medical Journal* 57: 373–87.

Hoerauf, Achim, and Ramakrishna U. Rao, eds. 2007. *Wolbachia: A Bug's Life in Another Bug*. Issues in Infectious Diseases. Volume 5. Basel, Switzerland: Karger Medical and Scientific Publishers.

Holmes, John C., and William M. Bethel. 1972. "Modification of Intermediate Host Behavior by Parasites." *Zoological Journal of the Linnean Society* 51 (S1): 123–49.

Hooper, John N. A. 2005. "Porifera (Sponges)." In *Marine Parasitology*, edited by Klaus Rohde, 174–77, 478–79. Melbourne, Australia: CSIRO Publishing.

Hornaday, William T. 1905. *Taxidermy and Zoological Collecting: A Complete Handbook for the Amateur Taxidermist, Collector, Osteologist, Museum-Builder, Sportsman, and Traveller*. Eighth Edition. New York: Charles Scribner's Sons.

Horne, Jon S., Edward O. Garton, and Janet L. Rachlow. 2008. "A Synoptic Model of Animal Space Use: Simultaneous Estimation of Home Range, Habitat Selection, and Inter/Intra-Specific Relationships." *Ecological Modelling* 214 (2): 338–48. https://doi.org/10.1016/j.ecolmodel.2008.02.042.

Hugot, Jean-Pierre, and Scott L. Gardner. 2000. "*Helminthoxys abrocomae* n. sp. (Nematoda: Oxyurida) from *Abrocoma cinerea* in Bolivia." *Systematic Parasitology* 47 (3): 223–30. https://doi.org/10.1023/A:1006460804935.

Hugot, Jean-Pierre, Scott L. Gardner, Victor Borba, Priscilla Araujo, Daniela Leles, Átila Augusto Stock Da-Rosa, Juliana Dutra, Luiz Fernando Ferreira, and Adauto Araújo. 2014. "Discovery of a 240 Million Year Old Nematode Parasite Egg in a Cynodont Coprolite Sheds Light on the Early Origin of Pinworms in Vertebrates." *Parasites & Vectors* 7 (486). https://doi.org/10.1186/s13071-014-0486-6.

Hugot, Jean-Pierre, Karl J. Reinhard, Scott L. Gardner, and Serge Morand. 1999. "Human Enterobiasis in Evolution: Origin, Specificity and Transmission." *Parasite* 6 (3): 201–8. https://doi.org/10.1051/parasite/1999063201.

Hunter, Philip. 2018. "The Revival of the Extended Phenotype." *EMBO Reports* 19 (7): e46477. https://doi.org/10.15252/embr.201846477.

Hurtado, A. Magdalena, M. Anderson Frey, Inés Hurtado, K. R. Hill, and Jack Baker. 2008. "The Role of Helminthes in Human Evolution." In *Medicine and Evolution: Current Applications, Future Prospects*, edited by Sarah Elton and Paul O'Higgins, 153–80. United States: CRC Press.

Hutterer, Rainer. 2001. "Diversity of Mammals in Bolivia." In *Biodiversity: A Challenge for Development Research and Policy*, edited by Wilhelm Barthlott, Matthias Winiger, and Nadja Biedinger, 279–88. Berlin: Springer. https://doi.org/10.1007/978-3-662-06071-1_18.

Hygnstrom, Scott E., Robert M. Timm, and Gary Eugene Larson. 1994. *Prevention and Control of Wildlife Damage*. University of Nebraska Cooperative Extension, Institute of Agriculture and Natural Resources, University of Nebraska–Lincoln.

Iarotski, Lev S., and Andrew Davis. 1981. "The Schistosomiasis Problem in the World: Results of a WHO Questionnaire Survey." *Bulletin of the World Health Organization* 59 (1): 115–27.

Ito, Akira, Gantigmaa Chuluunbaatar, Tetsuya Yanagida, Anu Davaasuren, Battulga Sumiya, Mitsuhiko Asakawa, Toshiaki Ki, et al. 2013. "*Echinococcus* Species from Red Foxes, Corsac Foxes, and Wolves in Mongolia." *Parasitology* 140 (13): 1648–54. https://doi.org/10.1017/S0031182013001030.

Ito, Akira, and Christine M. Budke. 2015. "The Present Situation of Echinococcoses in Mongolia." *Journal of Helminthology* 89 (6): 680–88. https://doi.org/10.1017/S0022149X15000620.

Jabbar, Abdul, Ian Beveridge, and Malcolm S. Bryant. 2015. "Morphological and Molecular Observations on the Status of *Crassicauda magna*, a Parasite of the Subcutaneous Tissues of the Pygmy Sperm Whale, with a Re-evaluation of the Systematic Relationships of the Genus *Crassicauda*." *Parasitology Research* 114 (3): 835–41. https://doi.org/10.1007/s00436-014-4245-6.

Jangoux, Michel. 1984. "Diseases of Echinoderms." *Helgoländer Meeresuntersuchungen* 37 (1–4): 207–16.

Janovy Jr., John. 2002. "Concurrent Infections and the Community Ecology of Helminth Parasites." *Journal of Parasitology* 88 (3): 440–45. https://doi.org/10.1645/0022-3395(2002)088 [0440:CIATCE]2.0.CO;2.

Janovy Jr., John, Richard E. Clopton, David A. Clopton, Scott D. Snyder, Aris Efting, and Laura Krebs. 1995. "Species Density Distributions as Null Models for Ecologically Significant Interactions of Parasite Species in an Assemblage." *Ecological Modelling* 77 (2): 189–96. https://doi.org/10.1016/0304-3800(93)E0087-J.

Janovy Jr., John, Richard E. Clopton, and Tamara J. Percival. 1992. "The Roles of Ecological and Evolutionary Influences in Providing Structure to Parasite Species Assemblages." *Journal of Parasitology* 78 (4): 630–40. https://doi.org/10.2307/3283537.

Janz, Niklas, and Sören Nylin. 2008. "The Oscillation Hypothesis of Host-Plant Range and Speciation." In *Specialization, Speciation, and Radiation: The Evolutionary Biology of Herbivorous Insects*, edited by Kelley J. Tilmon, 203–15. University of California Press.

Janzen, Daniel H. 1985. "On Ecological Fitting." *Oikos* 45 (3): 308–10.

Jenkins, Emily J., Louisa J. Castrodale, Simone J. C. de Rosemond, Brent R. Dixon, Stacey A. Elmore, Karen M. Gesy, Eric P. Hoberg, et al. 2013. "Tradition and Transition: Parasitic Zoonoses of People and Animals in Alaska, Northern Canada, and Greenland." In *Advances in Parasitology* 82: 33–204. Academic Press.

Jiménez, F. Agustín, Janet K. Braun, Mariel L. Campbell, and Scott L. Gardner. 2008. "Endoparasites of Fat-Tailed Mouse Opossums (*Thylamys didelphidae*) from Northwestern Argentina and Southern Bolivia, with the Description of a New Species of Tapeworm." *Journal of Parasitology* 94 (5): 1098–1102. https://doi.org/10.1645/GE-1424.1.

Jiménez-Uzcátegui, Gustavo, Bryan Milstead, Cruz Márquez, Javier Zabala, Paola Buitrón, Alizon Llerena, Sandie Salazar, and Birgit Fessl. 2006. "Galapagos Vertebrates: Endangered Status and Conservation Actions." *Galapagos Report* 2006–2007.

Johnson, Karl M. 2001. "Zoonotic Diseases—An Interview with Karl M. Johnson, MD by Vicky Glaser." *Vector-Borne and Zoonotic Diseases* 1 (3): 243–48. https://doi.org/10.1089/15303 6601753552611.

Johnston, T. Harvey, and Patricia M. Mawson. 1939. "Internal Parasites of the Pigmy Sperm Whale." *Records of the South Australian Museum* 6 (3): 263–74.

Jones, Arlene, Rodney A. Bray, and David I. Gibson, eds. 2005. *Keys to the Trematoda. Volume 2.* Wallingford: CABI.

Jones, Arlene, Lotfi F. Khalil, and Rodney A. Bray, eds. 1994. *Keys to the Cestode Parasites of Vertebrates.* CAB International.

Kajihara, Noriaki, and Kenji Hirayama. 2011. "The War against a Regional Disease in Japan: A History of the Eradication of *Schistosomiasis japonica*." *Tropical Medicine and Health* 39 (Suppl 1): 3.

Kallio, Eva R., Michael Begon, Heikki Henttonen, Esa Koskela, Tapio Mappes, Antti Vaheri, and Olli Vapalahti. 2009. "Cyclic Hantavirus Epidemics in Humans—Predicted by Rodent Host Dynamics." *Epidemics* 1 (2): 101–7. https://doi.org/10.1016/j.epidem.2009.03.002.

Kalogianni, Eleni, Nikol Kmentová, Eileen Harris, Brian Zimmerman, Sofia Giakoumi, Yorgos Chatzinikolaou, and Maarten P. M. Vanhove. 2017. "Occurrence and Effect of Trematode Metacercariae in Two Endangered Killifishes from Greece." *Parasitology Research* 116 (11): 3007–18. https://doi.org/10.1007/s00436-017-5610-z.

Kanev, Ivan. 1994. "Life-Cycle, Delimitation and Redescription of *Echinostoma revolutum* (Froelich, 1802) (Trematoda: Echinostomatidae)." *Systematic Parasitology* 28 (2): 125–44. https://doi.org/10.1007/BF00009591.

Kempema, Silka. 2007. "The Influence of Grazing Systems on Grassland Bird Density, Productivity, and Species Richness on Private Rangeland in the Nebraska Sandhills." Thesis. University of Nebraska–Lincoln.

Kennedy, Clive R. 1999. "Post-Cyclic Transmission in *Pomphorhynchus laevis* (Acanthocephala)." *Folia Parasitologica* 46 (2): 111–16.

Kennedy, Clive R., P. F. Broughton, and P. M. Hine. 1978. "The Status of Brown and Rainbow Trout, *Salmo trutta* and *S. gairdneri* as Hosts of the Acanthocephalan, *Pomphorhynchus laevis.*" *Journal of Fish Biology* 13 (2): 265–75. https://doi.org/10.1111/j.1095-8649.1978.tb03434.x.

Kiefer, Daniel, Michael Stubbe, Annegret Stubbe, Scott L. Gardner, D. Tserenorov, R. Samiya, D. Otgonbaatar, D. Sumiya, and Matthias S. Kiefer. 2012. "Siphonaptera of Mongolia and Tuva: Results of the Mongolian-German Biological Expeditions since 1962–Years 1999–2003." In *Erforschung Biologischer Ressourcen Der Mongolei*, 12: 153–67. Halle-Wittenberg: Institut für Biologie der Martin-Luther-Universität.

Kim, Myeong-Ju, Dong Hoon Shin, Mi-Jin Song, Hye-Young Song, and Min Seo. 2013. "Paleoparasitological Surveys for Detection of Helminth Eggs in Archaeological Sites of Jeolla-Do and Jeju-Do." *Korean Journal of Parasitology* 51 (4): 489.

Kinsella, John M., and Vasyl V. Tkach. 2009. "Checklist of Helminth Parasites of Soricomorpha (= Insectivora) of North America North of Mexico." *Zootaxa* 1969 (1): 36–58.

Klotz, Stephen A., Patricia L. Dorn, Mark Mosbacher, and Justin O. Schmidt. 2014. "Kissing Bugs in the United States: Risk for Vector-Borne Disease in Humans." *Environmental Health Insights* 8 (S2): 49–59.

Krupnik, Igor, Lars F. Krutak, Willis Walunga, Vera Metcalf, and Arctic Studies Center (National Museum of Natural History). 2002. *Akuzilleput Igaqullghet = Our Words Put to Paper: Sourcebook in St. Lawrence Island Heritage and History*. Washington, D.C.: Arctic Studies Center, National Museum of Natural History, Smithsonian Institution. http://archive.org/details/akuzilleputigaqu03krup.

Kuhn, Thomas, Jaime García-Màrquez, and Sven Klimpel. 2011. "Adaptive Radiation within Marine Anisakid Nematodes: A Zoogeographical Modeling of Cosmopolitan, Zoonotic Parasites." *PLOS ONE* 6 (12): e28642. https://doi.org/10.1371/journal.pone.0028642.

Kurtén, Björn. 1972. *The Age of Mammals*. New York: Columbia University Press.

Kutz, Susan J., Eric P. Hoberg, John Nagy, Lydden Polley, and Brett Elkin. 2004. "'Emerging' Parasitic Infections in Arctic Ungulates." *Integrative and Comparative Biology* 44 (2): 109–18. https://doi.org/10.1093/icb/44.2.109.

Kutz, Susan J., Eric P. Hoberg, Lydden Polley, and Emily J. Jenkins. 2005. "Global Warming Is Changing the Dynamics of Arctic Host–Parasite Systems." *Proceedings of the Royal Society B: Biological Sciences* 272 (1581): 2571–76. https://doi.org/10.1098/rspb.2005.3285.

Kuzelka, Robert D., and Charles Flowerday. 1993. *Flat Water: A History of Nebraska and Its Water.* Resource Report, No. 12. Conservation and Survey Division, Institute of Agriculture and Natural Resources, University of Nebraska–Lincoln.

Kuzmin, Yuriy, Vasyl V. Tkach, and Scott D. Snyder. 2003. "The Nematode Genus *Rhabdias* (Nematoda: Rhabdiasidae) from Amphibians and Reptiles of the Nearctic." *Comparative Parasitology* 70 (2): 101–14. https://doi.org/10.1654/4075.

Lafferty, Kevin D. 1993. "The Marine Snail, *Cerithidea californica*, Matures at Smaller Sizes Where Parasitism Is High." *Oikos* 68: 3–11.

Lafferty, Kevin D., Andrew P. Dobson, and Armand M. Kuris. 2006. "Parasites Dominate Food Web Links." *Proceedings of the National Academy of Sciences* 103 (30): 11211–16.

Lagrue, Clement, and Robert Poulin. 2007. "Life Cycle Abbreviation in the Trematode *Coitocaecum parvum*: Can Parasites Adjust to Variable Conditions?" *Journal of Evolutionary Biology* 20 (3): 1189–95.

Lambert, Christine R., Scott L. Gardner, and Donald W. Duszynski. 1988. "Coccidia (Apicomplexa: Eimeriidae) from the Subterranean Rodent *Ctenomys opimus* Wagner (Ctenomyidae) from Bolivia, South America." *Journal of Parasitology* 74 (6): 1018–22.

Lambshead, P. John D. 1993. "Recent Developments in Marine Benthic Biodiversity Research." *Oceanis* 19: 5–24.

Lambshead, P. John D., and Guy Boucher. 2003. "Marine Nematode Deep-Sea Biodiversity— Hyperdiverse or Hype?" *Journal of Biogeography* 30 (4): 475–85. https://doi.org/10.1046/j.1365 -2699.2003.00843.x.

Leles, Daniela, Scott L. Gardner, Karl J. Reinhard, Alena Iñiguez, and Adauto Araújo. 2012. "Are *Ascaris lumbricoides* and *Ascaris suum* a Single Species?" *Parasites & Vectors* 5 (42). https://doi .org/10.1186/1756-3305-5-42.

Lempereur, Laetitia, Morgan Delobelle, Marjan Doom, Jan Haelters, Etienne Levy, Bertrand Losson, and Thierry Jauniaux. 2017. "*Crassicauda boopis* in a Fin Whale (*Balaenoptera physalus*) Ship-Struck in the Eastern North Atlantic Ocean." *Parasitology Open* 3. https://doi.org/10 .1017/pao.2017.10.

Lessa, Enrique P. 1990. "Morphological Evolution of Subterranean Mammals: Integrating Structural, Functional, and Ecological Perspectives." *Progress in Clinical and Biological Research* 335: 211–30.

Lessa, Enrique P., and Joseph A. Cook. 1998. "The Molecular Phylogenetics of Tuco-Tucos (Genus: *Ctenomys*, Rodentia: Octodontidae) Suggests an Early Burst of Speciation." *Molecular Phylogenetics and Evolution* 9 (1): 88–99. https://doi.org/10.1006/mpev.1997.0445.

Lewis, Paul D. 1974. "Helminths of Terrestrial Molluscs in Nebraska. II. Life Cycle of *Leucochloridium variae* McIntosh, 1932 (Digenea: Leucochloridiidae)." *Journal of Parasitology* 60 (2): 251–55. https://doi.org/10.2307/3278459.

Lim, Boo Liat, and Donald Heyneman. 1965. "Host-Parasite Studies of *Angiostrongylus cantonensis* (Nematoda, Metastrongylidae) in Malaysian Rodents: Natural Infection of Rodents and Molluscs in Urban and Rural Areas of Central Malaya." *Annals of Tropical Medicine & Parasitology* 59 (4): 425–33. https://doi.org/10.1080/00034983.1965.11686328.

Littlewood, D. Timothy J., and Kenneth de Baets, eds. 2015. *Fossil Parasites. Advances in Parasitology*, vol. 90. Academic Press.

Littlewood, D. Timothy J., and Rodney A. Bray. 2014. *Interrelationships of the Platyhelminthes*. Boca Raton, FL: CRC Press.

Lockyer, Anne E., Catherine S. Jones, Leslie R. Noble, and David Rollinson. 2004. "Trematodes and Snails: An Intimate Association." *Canadian Journal of Zoology* 82 (2): 251–69.

Loker, Eric, and Bruce Hofkin. 2015. *Parasitology: A Conceptual Approach*. New York: Garland Science.

Loope, David B., and James B. Swinehart. 2000. "Thinking like a Dune Field: Geologic History in the Nebraska Sand Hills." *Great Plains Research* 10: 5–35.

Loope, David B., James B. Swinehart, and Jon P. Mason. 1995. "Dune-Dammed Paleovalleys of the Nebraska Sand Hills: Intrinsic versus Climatic Controls on the Accumulation of Lake and Marsh Sediments." *GSA Bulletin* 107 (4): 396–406. https://doi.org/10.1130/0016-7606(1995)107<0396:DDPOTN>2.3.CO;2.

Loope, Lloyd L., Ole Hamann, and Charles P. Stone. 1988. "Comparative Conservation Biology of Oceanic Archipelagoes: Hawaii and the Galápagos." *BioScience* 38 (4): 272–82. https://doi.org/10.2307/1310851.

Lurie-Weinberger, Mor N., and Uri Gophna. 2015. "Archaea in and on the Human Body: Health Implications and Future Directions." *PLOS Pathogens* 11 (6): e1004833. https://doi.org/10.1371/journal.ppat.1004833.

MacArthur, Robert Helmer, and Edward O. Wilson. 1967. *The Theory of Island Biogeography*. Princeton, NJ: Princeton University Press.

MacFadden, Bruce J., Yang Wang, Thure E. Cerling, and Federico Anaya. 1994. "South American Fossil Mammals and Carbon Isotopes: A 25 Million-Year Sequence from the Bolivian Andes." *Palaeogeography, Palaeoclimatology, Palaeoecology* 107 (3): 257–68. https://doi.org/10.1016/0031-0182(94)90098-1.

Mackenzie, Charles D., Mamoun M. Homeida, Adrian D. Hopkins, and Joni C. Lawrence. 2012. "Elimination of Onchocerciasis from Africa: Possible?" *Trends in Parasitology* 28 (1): 16–22. https://doi.org/10.1016/j.pt.2011.10.003.

Makarikov, Arseny A., and Vasyl V. Tkach. 2013. "Two New Species of *Hymenolepis* (Cestoda: Hymenolepididae) from Spalacidae and Muridae (Rodentia) from Eastern Palearctic." *Acta Parasitologica* 58 (1): 37–49. https://doi.org/10.2478/s11686-013-0115-0.

Manter, Harold W. 1955. "The Zoogeography of Trematodes of Marine Fishes." *Experimental Parasitology* 4 (1): 62–86. https://doi.org/10.1016/0014-4894(55)90024-2.

———. 1963. "The Zoogeographical Affinities of Trematodes of South American Freshwater Fishes." *Systematic Zoology* 12 (2): 45–70. https://doi.org/10.2307/2411621.

Marcer, Federica, Enrico Negrisolo, Giovanni Franzo, Cinzia Tessarin, Mario Pietrobelli, and Erica Marchiori. 2019. "Morphological and Molecular Characterization of Adults and Larvae of *Crassicauda* spp. (Nematoda: Spirurida) from Mediterranean Fin Whales *Balaenoptera physalus* (Linnaeus, 1758)." *International Journal for Parasitology: Parasites and Wildlife* 9: 258–65. https://doi.org/10.1016/j.ijppaw.2019.06.004.

Marcogliese, David J. 2008. "The Impact of Climate Change on the Parasites and Infectious Diseases of Aquatic Animals." *Revue Scientifique et Technique* 27 (2): 467–84.

Marcogliese, David J., and Judith Price. 1997. "The Paradox of Parasites." *Global Biodiversity* 7 (3): 7–15.

Margulis, Lynn. 1970. *Origin of Eukaryotic Cells: Evidence and Research Implications for a Theory of the Origin and Evolution of Microbial, Plant, and Animal Cells on the Precambrian Earth.* Yale University Press.

———. 1981. *Symbiosis in Cell Evolution: Life and Its Environment on the Early Earth.* San Francisco: W. H. Freeman.

Martin, Walter Edwin. 1950. "*Euhaplorchis californiensis* n.g., n. sp., Heterophyidae, Trematoda, with Notes on Its Life-Cycle." *Transactions of the American Microscopical Society* 69 (2): 194–209. https://doi.org/10.2307/3223410.

Mason, Joseph A., James B. Swinehart, Ronald J. Goble, and David B. Loope. 2004. "Late-Holocene Dune Activity Linked to Hydrological Drought, Nebraska Sand Hills, USA." *The Holocene* 14 (2): 209–17. https://doi.org/10.1191/0959683604hl677rp.

Mason, Joseph A., James B. Swinehart, Paul R. Hanson, David B. Loope, Ronald J. Goble, Xiaodong Miao, and Rebecca L. Schmeisser. 2011. "Late Pleistocene Dune Activity in the Central Great Plains, USA." Quaternary Science Reviews 30 (27): 3858–70. https://doi.org/10.1016/j.quascirev.2011.10.005.

Mauldin, Matthew, Jeffrey Doty, Yoshinori Nakazawa, Ginny Emerson, and Darin Carroll. 2016. "The Importance of Mammalogy, Infectious Disease Research, and Biosafety in the Field." *MANTER: Journal of Parasite Biodiversity* 3 (August): 1–9.

McAlpine, Donald F., Lurie D. Murison, and Eric P. Hoberg. 1997. "New Records for the Pygmy Sperm Whale, *Kogia breviceps* (Physeteridae) from Atlantic Canada with Notes on Diet and Parasites." *Marine Mammal Science* 13 (4): 701–4.

McCarraher, D. Bruce. 1960. "The Nebraska Sandhill Lakes: Their Characteristics and Fisheries Management Problems." In *Nebraska Game and Parks Commission—White Papers, Conference Presentations, & Manuscripts*, 7. Bassett, Nebraska: Nebraska Game, Forestation and Parks Commission.

McElwain, Andrew. 2019. "Are Parasites and Diseases Contributing to the Decline of Freshwater Mussels (Bivalvia, Unionida)?" *Freshwater Mollusk Biology and Conservation* 22 (2): 85–89.

McIntosh, Allen. 1932. "Some New Species of Trematode Worms of the Genus *Leucochloridium carus*, Parasitic in Birds from Northern Michigan, with a Key and Notes on Other Species of the Genus." *Journal of Parasitology* 19 (1): 32–53. https://doi.org/10.2307/3271429.

McIntosh, Charles Barron. 1996. *The Nebraska Sand Hills: The Human Landscape.* University of Nebraska Press.

McNulty, Samantha N., Andrew S. Mullin, Jefferson A. Vaughan, Vasyl V. Tkach, Gary J. Weil, and Peter U. Fischer. 2012. "Comparing the Mitochondrial Genomes of *Wolbachia*-Dependent and Independent Filarial Nematode Species." *BMC Genomics* 13 (145). https://doi.org/10.1186/1471-2164-13-145.

Mehlhorn, Heinz. 2008. *Encyclopedia of Parasitology. Volumes 1–2.* Springer Science & Business Media.

Miller, Melissa A., John M. Kinsella, Ray W. Snow, Malorie M. Hayes, Bryan G. Falk, Robert N. Reed, Frank J. Mazzotti, Craig Guyer, and Christina M. Romagosa. 2018. "Parasite Spillover: Indirect Effects of Invasive Burmese Pythons." *Ecology and Evolution* 8 (2): 830–40. https://doi.org/10.1002/ece3.3557.

Mitchell, Piers D. 2013. "The Origins of Human Parasites: Exploring the Evidence for Endoparasitism throughout Human Evolution." *International Journal of Paleopathology* 3 (3): 191–98. https://doi.org/10.1016/j.ijpp.2013.08.003.

Mitchell, Piers D., Evilena Anastasiou, and Danny Syon. 2011. "Human Intestinal Parasites in Crusader Acre: Evidence for Migration with Disease in the Medieval Period." *International Journal of Paleopathology* 1 (3): 132–37. https://doi.org/10.1016/j.ijpp.2011.10.005.

Monks, Scott, Víctor Rafael Zárate-Ramírez, and Griselda Pulido-Flores. 2005. "Helminths of Freshwater Fishes from the Metztitlán Canyon Reserve of the Biosphere, Hidalgo, Mexico." *Comparative Parasitology* 72 (2): 212–19. https://doi.org/10.1654/4139.

Moore, Janice. 1981. "Asexual Reproduction and Environmental Predictability in Cestodes (Cyclophyllidea: Taeniidae)." *Evolution* 35 (4): 723–41. https://doi.org/10.2307/2408243.

———. 1983. "Responses of an Avian Predator and Its Isopod Prey to an Acanthocephalan Parasite." *Ecology* 64 (5): 1000–1015. https://doi.org/10.2307/1937807.

Moore, Janice, Michael Freehling, and Nicholas J. Gotelli. 1994. "Altered Behavior in Two Species of Blattid Cockroaches Infected with *Moniliformis moniliformis* (Acanthocephala)." *Journal of Parasitology* 80 (2): 220–23. https://doi.org/10.2307/3283750.

Morand, Serge, and Boris R. Krasnov. 2010. *The Biogeography of Host-Parasite Interactions*. Oxford University Press.

Morand, Serge, Boris R. Krasnov, and D. Timothy J. Littlewood, eds. 2015. *Parasite Diversity and Diversification*. Cambridge University Press.

Morand, Serge, Boris R. Krasnov, and Robert Poulin. 2007. *Micromammals and Macroparasites: From Evolutionary Ecology to Management*. New York: Springer.

Morand, Serge, Pierre Legendre, Scott L. Gardner, and Jean-Pierre Hugot. 1996. "Body Size Evolution of Oxyurid (Nematoda) Parasites: The Role of Hosts." *Oecologia* 107 (2): 274–82. https://doi.org/10.1007/BF00327912.

Morgan, Jess A. T., Randall J. Dejong, Grace O. Adeoye, Ebenezer D. O. Ansa, Constança S. Barbosa, Philippe Brémond, Italo M. Cesari, et al. 2005. "Origin and Diversification of the Human Parasite *Schistosoma mansoni*." *Molecular Ecology* 14 (12): 3889–3902. https://doi.org/10.1111/j.1365-294X.2005.02709.x.

Muniz-Pereira, Luís C., Fabiano M. Vieira, and José L. Luque. 2009. "Checklist of Helminth Parasites of Threatened Vertebrate Species from Brazil." *Zootaxa* 2123 (1): 1–45. https://doi.org/10.11646/zootaxa.2123.1.1.

Myers, Thomas P. 1995. "Paleoindian Occupation of the Eastern Sand Hills." *Plains Anthropologist* 40 (151): 61–68.

Nadler, Steven A., Ramon A. Carreno, Hugo H. Mejía-Madrid, J. Ullberg, C. Pagan, R. Houston, and Jean-Pierre Hugot. 2007. "Molecular Phylogeny of Clade III Nematodes Reveals Multiple Origins of Tissue Parasitism." *Parasitology* 134 (10): 1421–42. https://doi.org/10.1017/S0031182007002880.

Nakao, Minoru, Antti Lavikainen, Tetsuya Yanagida, and Akira Ito. 2013. "Phylogenetic Systematics of the Genus *Echinococcus* (Cestoda: Taeniidae)." *International Journal for Parasitology*, Zoonoses Special Issue, 43 (12): 1017–29. https://doi.org/10.1016/j.ijpara.2013.06.002.

Navarro, Gonzalo, and Mabel Maldonado. 2002. *Geografía Ecológica de Bolivia: Vegetación y Ambientes Acuáticos*. Cochabamba, Bolivia: Centro de Ecología Simón I. Patiño, Departamento de Difusión.

Navone, Graciela T., Juliana Notarnicola, Santiago Nava, María del Rosario Robles, Carlos Galliari, and Marcela Lareschi. 2009. "Arthropods and Helminths Assemblage in Sigmodontine Rodents from Wetlands of the Rio de La Plata, Argentina." *Mastozoología Neotropical* 16 (1): 121–33.

Near, Thomas J. 2002. "Acanthocephalan Phylogeny and the Evolution of Parasitism." *Integrative and Comparative Biology* 42 (3): 668–77. https://doi.org/10.1093/icb/42.3.668.

Nelwan, Martin L. 2019. "Schistosomiasis: Life Cycle, Diagnosis, and Control." *Current Therapeutic Research* 91: 5–9.

Nichol, Stuart T., Christina F. Spiropoulou, Sergey Morzunov, Pierre E. Rollin, Thomas G. Ksiazek, Heinz Feldmann, Anthony Sanchez, James Childs, Sherif Zaki, and Clarence J. Peters. 1993. "Genetic Identification of a Hantavirus Associated with an Outbreak of Acute Respiratory Illness." *Science* 262 (5135): 914–17.

Niewiadomska, Katarzyna, and Teresa Pojmanska. 2011. "Multiple Strategies of Digenean Trematodes to Complete Their Life Cycles." *Wiadomości Parazytologiczne* 57 (4): 233–41.

Notarnicola, Juliana, F. Agustín Jiménez, and Scott L. Gardner. 2007. "A New Species of *Dipetalonema* (Filarioidea: Onchocercidae) from *Ateles chamek* from the Beni of Bolivia." *Journal of Parasitology* 93 (3): 661–67.

Notarnicola, Juliana, F. Agustín Jiménez Ruíz, and Scott L. Gardner. 2010. "*Litomosoides* (Nemata: Filarioidea) of Bats from Bolivia with Records for Three Known Species and the Description of a New Species." *Journal of Parasitology* 96 (4): 775–82. https://doi.org/10.1645/GE -2371.1.

Nozais, Jean-Pierre. 2003. "The Origin and Dispersion of Human Parasitic Diseases in the Old World (Africa, Europe and Madagascar)." *Memórias Do Instituto Oswaldo Cruz* 98 (Suppl. 1): 13–19. https://doi.org/10.1590/S0074-02762003000900004.

Nylin, Sören, Jessica Slove, and Niklas Janz. 2014. "Host Plant Utilization, Host Range Oscillations and Diversification in Nymphalid Butterflies: A Phylogenetic Investigation." *Evolution* 68 (1): 105–24. https://doi.org/10.1111/evo.12227.

Odum, Eugene P. 1971. *Fundamentals of Ecology*. 3rd ed. Philadelphia: W. B. Saunders Co. https://isbndb.com/book/9780721669410.

Oetinger, David F., and Brent B. Nickol. 1981. "Effects of Acanthocephalans on Pigmentation of Freshwater Isopods." *Journal of Parasitology* 67 (5): 672–84. https://doi.org/10.2307/3280441.

Olson, Peter D., Thomas H. Cribb, Vasyl V. Tkach, Rodney A. Bray, and D. Timothy J. Littlewood. 2003. "Phylogeny and Classification of the Digenea (Platyhelminthes: Trematoda)." *International Journal for Parasitology* 33 (7): 733–55. https://doi.org/10.1016/S0020-7519(03) 00049-3.

Olson, Peter D., and Vasyl V. Tkach. 2005. "Advances and Trends in the Molecular Systematics of the Parasitic Platyhelminthes." In *Advances in Parasitology* 60: 165–243. Academic Press. https://doi.org/10.1016/S0065-308X(05)60003-6.

Oh, Chang Seok, Min Seo, Jong Yil Chai, Sang Jun Lee, Myeung Ju Kim, Jun Bum Park, and Dong Hoon Shin. 2010. "Amplification and Sequencing of *Trichuris trichiura* Ancient DNA Extracted from Archaeological Sediments." *Journal of Archaeological Science* 37 (6): 1269–73. https://doi.org/10.1016/j.jas.2009.12.029.

Ōmura, Satoshi. 2016. "A Splendid Gift from the Earth: The Origins and Impact of the Avermectins (Nobel Lecture)." *Angewandte Chemie International Edition* 55 (35): 10190–209. https://doi.org/10.1002/anie.201602164.

Overstreet, Robin M., and Stephen S. Curran. 2004. "Defeating Diplostomoid Dangers in USA Catfish Aquaculture." *Folia Parasitologica* 51 (2–3): 153–65.

Palm, Harry W. 2011. "Fish Parasites as Biological Indicators in a Changing World: Can We Monitor Environmental Impact and Climate Change?" In *Progress in Parasitology*, edited by Heinz Mehlhorn, 223–50. Parasitology Research Monographs. Berlin: Springer. https://doi.org/10.1007/978-3-642-21396-0_12.

Parvate, Amar, Evan P. Williams, Mariah K. Taylor, Yong-Kyu Chu, Jason Lanman, Erica Ollmann Saphire, and Colleen B. Jonsson. 2019. "Diverse Morphology and Structural Features of Old and New World Hantaviruses." *Viruses* 11 (9): 862. https://doi.org/10.3390/v11090862.

Patton, James L, and Mark S. Hafner. 1983. "Biosystematics of the Native Rodents of the Galapagos Archipelago, Ecuador." In *Patterns of Evolution in Galapagos Organisms*, edited by Robert I. Bowman, Margaret Berson, and Alan E. Leviton, 539–68. San Francisco: AAAS.

Pavesi, Angelo. 2005. "Microbes Coevolving with Human Host and Ancient Human Migrations." *Journal of Anthropological Sciences* 83: 9–28.

Pennisi, Elizabeth. 2019. "DNA Barcodes Jump-Start Search for New Species." *Science* 364 (6444): 920–21. https://doi.org/10.1126/science.364.6444.920.

Pellis, Sergio M., Tamara J. Pasztor, Vivien C. Pellis, and Donald A. Dewsbury. 2000. "The Organization of Play Fighting in the Grasshopper Mouse (*Onychomys leucogaster*): Mixing Predatory and Sociosexual Targets and Tactics." *Aggressive Behavior* 26 (4): 319–34. https://doi.org/10.1002/1098-2337(2000)26:4<319::AID-AB4>3.0.CO;2-I.

Pfeiffer, Kent E., and Allen A. Steuter. 1994. "Preliminary Response of Sandhills Prairie to Fire and Bison Grazing." *Rangeland Ecology & Management / Journal of Range Management Archives* 47 (5): 395–97.

Plyusnina, Angelina, Emöke Ferenczi, Gabor R. Racz, Kirill Nemirov, Åke Lundkvist, Antti Vaheri, Olli Vapalahti, and Alexander Plyusnin. 2009. "Co-circulation of Three Pathogenic Hantaviruses: Puumala, Dobrava, and Saaremaa in Hungary." *Journal of Medical Virology* 81 (12): 2045–52. https://doi.org/10.1002/jmv.21635.

Poinar Jr., George, and A. J. Boucot. 2006. "Evidence of Intestinal Parasites of Dinosaurs." *Parasitology* 133 (2): 245–49. https://doi.org/10.1017/S0031182006000138.

Poinar Jr., George, and Roberta Poinar. 2010. *What Bugged the Dinosaurs? Insects, Disease, and Death in the Cretaceous*. Princeton University Press.

Polley, Lydden, Eric P. Hoberg, and Susan J. Kutz. 2010. "Climate Change, Parasites and Shifting Boundaries." *Acta Veterinaria Scandinavica* 52 (1): S1. https://doi.org/10.1186/1751 -0147-52-S1-S1.

Ponton, Fleur, Fernando Otálora-Luna, Thierry Lefèvre, Patrick M. Guerin, Camille Lebarbenchon, David Duneau, David G. Biron, and Frédéric Thomas. 2011. "Water-Seeking Behavior in Worm-Infected Crickets and Reversibility of Parasitic Manipulation." *Behavioral Ecology* 22 (2): 392–400. https://doi.org/10.1093/beheco/arq215.

Poulin, Robert, Jacques Brodeur, and Janice Moore. 1994. "Parasite Manipulation of Host Behaviour: Should Hosts Always Lose?" *Oikos* 70 (3): 479–84. https://doi.org/10.2307/3545788.

Poulin, Robert, Megan Wise, and Janice Moore. 2003. "A Comparative Analysis of Adult Body Size and Its Correlates in Acanthocephalan Parasites." *International Journal for Parasitology* 33 (8): 799–805. https://doi.org/10.1016/S0020-7519(03)00108-5.

Pritchard, Mary Hanson, and Günther O.W. Kruse. 1982. *The Collection and Preservation of Animal Parasites.* Lincoln: University of Nebraska Press.

Pulido-Flores, Griselda, and Scott Monks. 2005. "Monogenean Parasites of Some Elasmobranchs (Chondrichthyes) from the Yucatán Peninsula, Mexico." *Comparative Parasitology* 72 (1): 69–74. https://doi.org/10.1654/4049.

Racz, Gabor R., Enikő Bán, Emőke Ferenczi, and György Berencsi. 2006. "A Simple Spatial Model to Explain the Distribution of Human Tick-Borne Encephalitis Cases in Hungary." *Vector-Borne and Zoonotic Diseases* 6 (4): 369–78. https://doi.org/10.1089/vbz.2006.6.369.

Ratcliffe, Brett C. 1998. "Insects." In *An Atlas of the Sand Hills*, 3rd ed., edited by Ann Salomon Bleed and Charles Flowerday, 143–54. Resource Atlas, no. 5b. Lincoln: Conservation and Survey Division, Institute of Agriculture and Natural Resources, University of Nebraska–Lincoln.

Rausch, Robert L. 1952. "Studies on the Helminth Fauna of Alaska. XI. Helminth Parasites of Microtine Rodents: Taxonomic Considerations." *Journal of Parasitology* 38 (5): 415–44. https:// doi.org/10.2307/3273922.

———. 1953a. "On the Land Mammals of St. Lawrence Island, Alaska." *The Murrelet* 34 (2): 18–26. https://doi.org/10.2307/3535866.

———. 1953b. "Studies on the Helminth Fauna of Alaska. XIII. Disease in the Sea Otter, with Special Reference to Helminth Parasites." *Ecology* 34 (3): 584–604. https://doi.org/10.2307 /1929729.

———. 1953c. "The Taxonomic Value and Variability of Certain Structures in the Cestode Genus *Echinococcus* (Rudolphi, 1801) and a Review of Recognized Species." In *Thapar Commemoration Volume 1953: A Collection of Articles Presented to Prof. G. S. Thapar on His 60th Birthday*, edited by Jagdeshwari Dayal.

———. 1975. "Cestodes of the Genus *Hymenolepis* Weinland, 1858 (sensu lato) from Bats in North America and Hawaii." *Canadian Journal of Zoology* 53 (11): 1537–51. https://doi.org /10.1139/z75-189.

Rausch, Robert L., and Everett L. Schiller. 1956. "Studies on the Helminth Fauna of Alaska: XXV. The Ecology and Public Health Significance of *Echinococcus sibiricensis* Rausch & Schiller, 1954, on St Lawrence Island." *Parasitology* 46 (3–4): 395–419. https://doi.org/10.1017/S0031182 000026561.

Rausch, Robert L., and Francis S. L. Williamson. 1959. "Studies on the Helminth Fauna of Alaska. XXXIV. The Parasites of Wolves, *Canis lupus* L." *Journal of Parasitology* 45 (4): 395–403. https://doi.org/10.2307/3274390.

Rausch, Robert L., and Stephen H. Richards. 1971. "Observations on Parasite–Host Relationships of *Echinococcus multilocularis* Leuckart, 1863, in North Dakota." *Canadian Journal of Zoology* 49 (10): 1317–30. https://doi.org/10.1139/z71-198.

Rausch, Robert L., F. H. Fay, and Francis S. L. Williamson. 1990. "The Ecology of *Echinococcus multilocularis* (Cestoda: Taeniidae) on St. Lawrence Island, Alaska.—II.—Helminth Populations in the Definitive Host." *Annales de Parasitologie Humaine et Comparée* 65 (3): 131–40. https://doi.org/10.1051/parasite/1990653131.

Ravasi, Damiana F., Mannus J. O'Riain, Faezah Davids, and Nicola Illing. 2012. "Phylogenetic Evidence That Two Distinct Trichuris Genotypes Infect Both Humans and Non-Human Primates." *PLOS ONE* 7 (8): p.e44187. https://doi.org/10.1371/journal.pone.0044187.

Raven, Peter H. 2002. "Science, Sustainability, and the Human Prospect." *Science* 297 (5583): 954–58. https://doi.org/10.1126/science.297.5583.954.

Raven, Peter H., and Daniel I. Axelrod. 1974. "Angiosperm Biogeography and Past Continental Movements." *Annals of the Missouri Botanical Garden* 61 (3): 539–673. https://doi.org/10.2307/2395021.

Raven, Peter H., and Scott E. Miller. 2020. "Here Today, Gone Tomorrow." *Science* 370 (6513): 149. https://doi.org/10.1126/science.abf1185.

Reichman, O. James, Thomas G. Whitham, and George A. Ruffner. 1982. "Adaptive Geometry of Burrow Spacing in Two Pocket Gopher Populations." *Ecology* 63 (3): 687–95. https://doi.org/10.2307/1936789.

Resetarits, Emlyn J., Mark E. Torchin, and Ryan F. Hechinger. 2020. "Social Trematode Parasites Increase Standing Army Size in Areas of Greater Invasion Threat." *Biology Letters* 16 (2): 1–7. https://doi.org/10.1098/rsbl.2019.0765.

Richards, Charles S. 1977. "*Schistosoma mansoni*: Susceptibility Reversal with Age in the Snail Host *Biomphalaria glabrata*." *Experimental Parasitology* 42 (1): 165–68. https://doi.org/10.1016/0014-4894(77)90074-1.

Richards, Frank, Donald Hopkins, and Ed Cupp. 2000. "Programmatic Goals and Approaches to Onchocerciasis." *Lancet* 355 (9216): 1663–64.

Ricklefs, Robert E. 1973. *Ecology*. 2nd ed. Newton, MS: Chiron Press.

Ripperger, Simon P., Sebastian Stockmaier, and Gerald G. Carter. 2020. "Tracking Sickness Effects on Social Encounters via Continuous Proximity Sensing in Wild Vampire Bats." *Behavioral Ecology* 31 (6): 1296–1302. https://doi.org/10.1093/beheco/araa111.

Roberts, Larry S., and John Janovy. 2000. *Gerald D. Schmidt & Larry S. Roberts' Foundations of Parasitology*. 6th ed. Boston: McGraw Hill.

Robles, María del Rosario, Graciela T. Navone, and Juliana Notarnicola. 2006. "A New Species of *Trichuris* (Nematoda: Trichuridae) from Phyllotini Rodents in Argentina." *Journal of Parasitology* 92 (1): 100–104. https://doi.org/10.1645/GE-GE-552R.1.

Rodrigues, Priscila T., Hugo O. Valdivia, Thais C. de Oliveira, João Marcelo P. Alves, Ana Maria R. C. Duarte, Crispim Cerutti-Junior, Julyana C. Buery, et al. 2018. "Human Migration and the Spread of Malaria Parasites to the New World." *Scientific Reports* 8 (1): 1993. https://doi.org/10.1038/s41598-018-19554-0.

Rossin, Alejandra, and Ana I. Malizia. 2002. "Relationship between Helminth Parasites and Demographic Attributes of a Population of the Subterranean Rodent *Ctenomys talarum* (Rodentia: Octodontidae)." Journal of Parasitology 88 (6): 1268–70. https://doi.org/10.1645/0022-3395(2002)088[1268:RBHPAD]2.0.CO;2.

Rostami, A., S. M. Riahi, H. R. Gamble, Y. Fakhri, M. Nourollahpour Shiadeh, M. Danesh, H. Behniafar, et al. 2020. "Global Prevalence of Latent Toxoplasmosis in Pregnant Women: A Systematic Review and Meta-Analysis." *Clinical Microbiology and Infection* 26 (6): 673–83. https://doi.org/10.1016/j.cmi.2020.01.008.

Rueppell, Olav, Miranda K. Hayworth, and N. P. Ross. 2010. "Altruistic Self-Removal of Health-Compromised Honey Bee Workers from Their Hive." *Journal of Evolutionary Biology* 23 (7): 1538–46. https://doi.org/10.1111/j.1420-9101.2010.02022.x.

Ruiz, Gregory M., and David R. Lindberg. 1989. "A Fossil Record for Trematodes: Extent and Potential Uses." *Lethaia* 22 (4): 431–38. https://doi.org/10.1111/j.1502-3931.1989.tb01447.x.

Russo, Isa-Rita M., Sean Hoban, Paulette Bloomer, Antoinette Kotzé, Gernot Segelbacher, Ian Rushworth, Coral Birss, and Michael W. Bruford. 2019. "'Intentional Genetic Manipulation' as a Conservation Threat." *Conservation Genetics Resources* 11 (2): 237–47. https://doi.org/10.1007/s12686-018-0983-6.

Sage, Richard D., Donald Heyneman, Kee-Chong Lim, and Allan C. Wilson. 1986. "Wormy Mice in a Hybrid Zone." *Nature* 324 (6092): 60–63. https://doi.org/10.1038/324060a0.

Salm, Andrea, and Jürg Gertsch. 2019. "Cultural Perception of Triatomine Bugs and Chagas Disease in Bolivia: A Cross-Sectional Field Study." *Parasites & Vectors* 12 (1): 291. https://doi.org/10.1186/s13071-019-3546-0.

Sato, Hiroshi, Munehiro Okamoto, Masashi Ohbayashi, and Maria Gloria Basanez. 1988. "A New Cestode, *Raillietina (Raillietina) oligocapsulata* n. sp., and *R. (R.) demerariensis* (Daniels, 1895) from Venezuelan Mammals." *Japanese Journal of Veterinary Research* 36 (1): 31–45.

Schacht, Walter H., Jerry D. Volesky, Dennis Bauer, Alexander Smart, and Eric Mousel. 2000. "Plant Community Patterns on Upland Prairie in the Eastern Nebraska Sandhills." *Prairie Naturalist* 32 (1): 43–58.

Schell, Stewart Claude. 1970. *How to Know the Trematodes*. W. C. Brown Company.

Schmidt, Gerald D. 1970. *How to Know the Tapeworms*. W. C. Brown Company.

———. 1986. *CRC Handbook of Tapeworm Identification*. Boca Raton, FL: CRC-Press.

Schmidt, Gerald D., and Larry S. Roberts. 1977. *Foundations of Parasitology*. Mosby.

Schmieder, Jens, Sherilyn C. Fritz, James B. Swinehart, Avery L. C. Shinneman, Alexander P. Wolfe, Gifford Miller, N. Daniels, K. C. Jacobs, and Eric C. Grimm. 2011. "A Regional-Scale Climate Reconstruction of the Last 4000 Years from Lakes in the Nebraska Sand Hills, USA." *Quaternary Science Reviews* 30 (13–14): 1797–1812.

Schmeisser, Rebecca L., David B. Loope, and David A. Wedin. 2009. "Clues to the Medieval Destabilization of the Nebraska Sand Hills, USA, from Ancient Pocket Gopher Burrows." *PALAIOS* 24 (12): 809–17. https://doi.org/10.2110/palo.2009.p09-037r.

Schmunis, Gabriel A. 2007. "Epidemiology of Chagas Disease in Non Endemic Countries: The Role of International Migration." *Memórias Do Instituto Oswaldo Cruz* 102 (October): 75–86. https://doi.org/10.1590/S0074-02762007005000093.

Scholz, Tomáš, Roman Kuchta, and Jan Brabec. 2019. "Broad Tapeworms (Diphyllobothriidae), Parasites of Wildlife and Humans: Recent Progress and Future Challenges." *International Journal for Parasitology: Parasites and Wildlife* 9: 359–69. https://doi.org/10.1016/j.ijppaw.2019.02.001.

Sheng, Jinliang, Mengmeng Jiang, Meihua Yang, Xinwen Bo, Shanshan Zhao, Yanyan Zhang, Hazihan Wureli, Baoju Wang, Changchun Tu, and Yuanzhi Wang. 2019. "Tick Distribution in Border Regions of Northwestern China." *Ticks and Tick-Borne Diseases* 10 (3): 665–69. https://doi.org/10.1016/j.ttbdis.2019.02.011.

Simpson, George Gaylord. 1980. *Splendid Isolation: The Curious History of South American Mammals.* Yale University Press.

Skinner, John D., and Christian T. Chimimba. 2005. *The Mammals of the Southern African Subregion.* Cambridge University Press.

Skryabin, A. S. 1961. "*Tetragonoporus calyptocephalus* n.g., n.sp. from the Sperm Whale." *Helminthologia* 3 (1/4): 311–15.

———. 1967. "*Polygonoporus giganticus* n.g., n.sp., a Parasite of Sperm Whales." *Parazitologiya* 1 (2): 131–36.

Skuce, Philip J., Eric R. Morgan, Jan van Dijk, and Malcolm Mitchell. 2013. "Animal Health Aspects of Adaptation to Climate Change: Beating the Heat and Parasites in a Warming Europe." *Animal* 7 (S2): 333–45. https://doi.org/10.1017/S175173111300075X.

Smit, Nico J., Niel L. Bruce, and Kerry A. Hadfield, eds. 2019. *Parasitic Crustacea: State of Knowledge and Future Trends.* Vol. 3. Zoological Monographs. Cham, Switzerland: Springer International.

Smythe, Ashleigh B., Oleksandr Holovachov, and Kevin M. Kocot. 2019. "Improved Phylogenomic Sampling of Free-Living Nematodes Enhances Resolution of Higher-Level Nematode Phylogeny." *BMC Evolutionary Biology* 19 (1): 121. https://doi.org/10.1186/s12862-019-1444-x.

Sokolow, Susanne H., Chelsea L. Wood, Isabel J. Jones, Kevin D. Lafferty, Armand M. Kuris, Michael H. Hsieh, and Giulio A. De Leo. 2018. "To Reduce the Global Burden of Human Schistosomiasis, Use 'Old Fashioned' Snail Control." *Trends in Parasitology* 34 (1): 23–40.

Solari, Sergio, Víctor Pacheco, Lucía Luna, Paul M. Velazco, and Bruce D. Patterson. 2006. "Mammals of the Manu Biosphere Reserve." *Fieldiana Zoology* 2006 (110): 13–22.

Song, J.-W., L. J. Baek, J. W. Nagle, D. Schlitter, and R. Yanagihara. 1996. "Genetic and Phylogenetic Analyses of Hantaviral Sequences Amplified from Archival Tissues of Deer Mice (*Peromyscus maniculatus nubiterrae*) Captured in the Eastern United States." *Archives of Virology* 141 (5): 959–67. https://doi.org/10.1007/BF01718170.

Southwell, T., and Baini Prashad. 1918. "Methods of Asexual and Parthenogenetic Reproduction in Cestodes." *Journal of Parasitology* 4 (3): 122–29. https://doi.org/10.2307/3271029.

Steiner-Souza, Francisco, Thales R. O. De Freitas, and Pedro Cordeiro-Estrela. 2010. "Inferring Adaptation within Shape Diversity of the Humerus of Subterranean Rodent *Ctenomys.*"

Biological Journal of the Linnean Society 100 (2): 353–67. https://doi.org/10.1111/j.1095-8312.2010.01400.x.

Stevens, Jamie R., Harry A. Noyes, Gabriel A. Dover, and Wendy C. Gibson. 1999. "The Ancient and Divergent Origins of the Human Pathogenic Trypanosomes, *Trypanosoma brucei* and *T. cruzi*." *Parasitology* 118 (1): 107–16. https://doi.org/10.1017/S0031182098003473.

Stiles, Charles Wardell. 1939. "Early History, in Part Esoteric, of the Hookworm (Uncinariasis) Campaign in Our Southern United States." *Journal of Parasitology* 25 (4): 283–308.

Strickland, G. Thomas. 2006. "Liver Disease in Egypt: Hepatitis C Superseded Schistosomiasis as a Result of Iatrogenic and Biological Factors." *Hepatology* 43 (5): 915–22. https://doi.org/10.1002/hep.21173.

Stubbendieck, James, Theresa R. Flessner, and Ronald Weedon. 1989. "Blowouts in the Nebraska Sandhills: The Habitat of *Penstemon haydenii*." In *Prairie Pioneers: Ecology, History and Culture: Proceedings of the Eleventh North American Prairie Conference Held 7–11 August 1988, Lincoln, Nebraska,* 223–26.

Summers, Robert W., David E. Elliott, Khurram Qadir, Joseph F. Urban, Robin Thompson, and Joel V. Weinstock. 2003. "*Trichuris suis* Seems to Be Safe and Possibly Effective in the Treatment of Inflammatory Bowel Disease." *American Journal of Gastroenterology* 98 (9): 2034–41. https://doi.org/10.1016/S0002-9270(03)00623-3.

Swinehart, James B., and Robert F. Diffendal. 1989. "Geology of the Pre-Dune Strata." In *An Atlas of the Sand Hills.* Conservation and Survey Division, Institute of Agriculture and Natural Resources, University of Nebraska-Lincoln. https://digitalcommons.unl.edu/natrespapers/581.

Tain, Luke, Marie-Jeanne Perrot-Minnot, and Frank Cézilly. 2006. "Altered Host Behaviour and Brain Serotonergic Activity Caused by Acanthocephalans: Evidence for Specificity." *Proceedings of the Royal Society B: Biological Sciences* 273 (1605): 3039–45. https://doi.org/10.1098/rspb.2006.3618.

Tamarozzi, Francesca, Alice Halliday, Katrin Gentil, Achim Hoerauf, Eric Pearlman, and Mark J. Taylor. 2011. "Onchocerciasis: The Role of *Wolbachia* Bacterial Endosymbionts in Parasite Biology, Disease Pathogenesis, and Treatment." *Clinical Microbiology Reviews* 24 (3): 459–68. https://doi.org/10.1128/CMR.00057-10.

Tanaka, H., and Moriyasu Tsuji. 1997. "From Discovery to Eradication of Schistosomiasis in Japan: 1847–1996." *International Journal for Parasitology* 27 (12): 1465–80. https://doi.org/10.1016/S0020-7519(97)00183-5.

Taylor, Mark J., Helen F. Cross, and Katja Bilo. 2000. "Inflammatory Responses Induced by the Filarial Nematode *Brugia malayi* Are Mediated by Lipopolysaccharide-like Activity from Endosymbiotic Wolbachia Bacteria." *Journal of Experimental Medicine* 191 (8): 1429–36. https://doi.org/10.1084/jem.191.8.1429.

Telford Jr., Sam R. 2016. *Hemoparasites of the Reptilia: Color Atlas and Text.* CRC Press.

Telford Jr., Sam R., and Charles R. Bursey. 2003. "Comparative Parasitology of Squamate Reptiles Endemic to Scrub and Sandhills Communities of North-Central Florida, U.S.A." *Comparative Parasitology* 70 (2): 172–81. https://doi.org/10.1654/4060.

Tenora, František, and Éva Murai. 1975. "Cestodes Recovered from Rodents (Rodentia) in Mongolia." *Annales Historico-Naturales Musei Nationalis Hungarici* 67: 65–70.

Thomas, Frédéric, Shelley Adamo, and Janice Moore. 2005. "Parasitic Manipulation: Where Are We and Where Should We Go?" *Behavioural Processes* 68 (3): 185–99.

Thomas, Peter O. 1988. "Kelp Gulls, *Larus dominicanus*, Are Parasites on Flesh of the Right Whale, *Eubalaena australis*." *Ethology* 79 (2): 89–103. https://doi.org/10.1111/j.1439-0310.1988.tb00703.x.

Thompson, John N. 2005. *The Geographic Mosaic of Coevolution*. Chicago: University of Chicago Press.

Tinnin, David S., Jonathan L. Dunnum, Jorge Salazar-Bravo, Nyamsuren Batsaikhan, M. Scott Burt, Scott L. Gardner, and Terry L. Yates. 2002. "Contributions to the Mammalogy of Mongolia, with a Checklist of the Species of the Country." *Special Publications, Museum of Southwestern Biology* 6 (October): 1–38.

Tinnin, David S., Sumiya Ganzorig, and Scott L. Gardner. 2011a. "Helminths of Squirrels (Sciuridae) from Mongolia." *Occasional Papers Museum of Texas Tech University* 303 (October): 1–9.

———. 2011b. "Helminths of Small Mammals (Erinaceomorpha, Soricomorpha, Chiroptera, Rodentia, and Lagomorpha) of Mongolia." *Special Publications of the Museum of Texas Tech University* 59 (October): 1–50.

Tinnin, David S., Scott L. Gardner, and Sumiya Ganzorig. 2008. "Helminths of Small Mammals (Chiroptera, Insectivora, Lagomorpha) from Mongolia with a Description of a New Species of *Schizorchis* (Cestoda: Anoplocephalidae)." *Comparative Parasitology* 75 (1): 107–14. https://doi.org/10.1654/4288.1.

Tinnin, David S., Ethan T. Jensen, Nyamsuren Batsaikhan, and Scott L. Gardner. 2012. "Coccidia (Apicomplexa: Eimeriidae) from *Vespertilio murinus* and *Eptesicus gobiensis* (Chiroptera: Vespertilionidae) in Mongolia and How Many Species of Coccidia Occur in Bats?" *Erforschung Biologischer Ressourcen Der Mongolei* 12 (January): 117–24.

Tkach, Vasyl V., Jay A. Schroeder, Stephen E. Greiman, and Jefferson A. Vaughan. 2012. "New Genetic Lineages, Host Associations and Circulation Pathways of *Neorickettsia* Endosymbionts of Digeneans." *Acta Parasitologica* 57 (3): 285–92. https://doi.org/10.2478/s11686-012-0043-4.

Toledo, Rafael, Valentin Radev, Ivan Kanev, Scott L. Gardner, and Bernard Fried. 2014. "History of Echinostomes (Trematoda)." *Acta Parasitologica* 59 (4): 555–67. https://doi.org/10.2478/s11686-014-0302-7.

Triantis, Kostas A., and Thomas J. Matthews. 2020. "Biodiversity Theory Backed by Island Bird Data." *Nature* 579 (7797): 36–37. https://doi.org/10.1038/d41586-020-00426-5.

Tsai, Isheng J., Magdalena Zarowiecki, Nancy Holroyd, Alejandro Garciarrubio, Alejandro Sanchez-Flores, Karen L. Brooks, Alan Tracey, et al. 2013. "The Genomes of Four Tapeworm Species Reveal Adaptations to Parasitism." *Nature* 496 (7443): 57–63. https://doi.org/10.1038/nature12031.

Tufts, Danielle, Nyamsuren Batsaikhan, Michael Pitner, Gábor R. Rácz, Altangerel Dursahinhan, and Scott L. Gardner. 2016. "Identification of *Taenia* Metacestodes from Mongolian Mammals

Using Multivariate Morphometrics of the Rostellar Hooks." *Erforschung Biologischer Ressourcen Der Mongolei* 13 (January): 361–75.

Tyson, Edward. 1683a. "*Lumbricus latus*, or a Discourse Read before the Royal Society of the Joynted Worm, Wherein a Great Many Mistakes of Former Writers Concerning It, Are Remarked; Its Natural History from More Exact Observations Is Attempted; and the Whole Urged, as a Difficulty against the Doctrine of Univocal Generation." *Philosophical Transactions of the Royal Society of London* 13 (146): 113–44.

———. 1683b. "*Lumbricus teres*, or Some Anatomical Observations on the Round Worm Bred in Human Bodies." *Philosophical Transactions of the Royal Society of London* 13 (147): 154–61.

———. 1686. "*Lumbricus hydropicus*; Or an Essay to Prove That Hydatides Often Met with in Morbid Animal Bodies, Are a Species of Worms, or Imperfect Animals. By That Learned and Curious Anatomist Edward Tyson, MD and R. Soc. S." *Philosophical Transactions (1683–1775)* 16: 506–10.

University of Utah. 2019. "Early Humans Evolved in Ecosystems unlike Any Found Today." October 7, 2019. https://phys.org/news/2019-10-early-humans-evolved-ecosystems-today.html.

van den Hoogen, Johan, Stefan Geisen, Devin Routh, Howard Ferris, Walter Traunspurger, David A. Wardle, Ron G. M. de Goede, et al. 2019. "Soil Nematode Abundance and Functional Group Composition at a Global Scale." *Nature* 572 (7768): 194–98. https://doi.org/10.1038/s41586-019-1418-6.

Vapalahti, Olli, Jukka Mustonen, Åke Lundkvist, Heikki Henttonen, Alexander Plyusnin, and Antti Vaheri. 2003. "Hantavirus Infections in Europe." *The Lancet Infectious Diseases* 3 (10): 653–61. https://doi.org/10.1016/S1473-3099(03)00774-6.

Walker, Ernest P, and John L Paradiso. 1975. *Mammals of the World*. 3rd ed. Baltimore: Johns Hopkins University Press.

Wang, Shuai, Sen Wang, Yingfeng Luo, Lihua Xiao, Xuenong Luo, Shenghan Gao, Yongxi Dou, et al. 2016. "Comparative Genomics Reveals Adaptive Evolution of Asian Tapeworm in Switching to a New Intermediate Host." *Nature Communications* 7 (1): 12845. https://doi.org/10.1038/ncomms12845.

Weeks, Andrew R., Michael Turelli, William R. Harcombe, K. Tracy Reynolds, and Ary A. Hoffmann. 2007. "From Parasite to Mutualist: Rapid Evolution of *Wolbachia* in Natural Populations of *Drosophila*." *PLOS Biology* 5 (5): e114. https://doi.org/10.1371/journal.pbio.0050114.

Welker, Frido, Matthew J. Collins, Jessica A. Thomas, Marc Wadsley, Selina Brace, Enrico Cappellini, Samuel T. Turvey, et al. 2015. "Ancient Proteins Resolve the Evolutionary History of Darwin's South American Ungulates." *Nature* 522 (7554): 81–84. https://doi.org/10.1038/nature14249.

Werren, John H., Laura Baldo, and Michael E. Clark. 2008. "*Wolbachia*: Master Manipulators of Invertebrate Biology." *Nature Reviews Microbiology* 6 (10): 741–51. https://doi.org/10.1038/nrmicro1969.

Whitcomb, Robert F. 1989. "Nebraska Sand Hills: The Last Prairie." In *Proceedings of the Eleventh North American Prairie Conference—Prairie Pioneers: Ecology, History, and Culture*, edited by Thomas Bragg B. and James Stubbendieck, 57–69. Lincoln: University of Nebraska Printing.

Whitfield, John. 2001. "Humans and Tapeworm: A Long Story." *Nature*, April. https://doi.org/10
.1038/news010404-12.

Whitfield, P. J., and N. A. Evans. 1983. "Parthenogenesis and Asexual Multiplication among Parasitic Platyhelminths." *Parasitology* 86 (4): 121–60. https://doi.org/10.1017/S0031182000050873.

WHO. 2019. *World Malaria Report 2018*. Geneva: World Health Organization. https://www.who
.int/malaria/publications/world-malaria-report-2018/report/en/.

Wilkins, Kenneth T., and Heather R. Roberts. 2007. "Comparative Analysis of Burrow Systems of Seven Species of Pocket Gophers (Rodentia: Geomyidae)." *Southwestern Naturalist* 52 (1): 83–88.

Wilson, Don E., and DeeAnn M. Reeder, eds. 2005. *Mammal Species of the World: A Taxonomic and Geographic Reference*. 3rd ed. Baltimore: Johns Hopkins University Press.

Wilson, Edward O. 1985. "The Biological Diversity Crisis." *BioScience* 35 (11): 700–706.

———. 1999. *The Diversity of Life*. W. W. Norton & Company.

———. 2002. *The Future of Life*. Knopf Doubleday Publishing Group.

Wilson, Edward O., and G. Evelyn Hutchinson. 1989. "Robert Helmer MacArthur 1930–1972." In *Biographical Memoirs. Volume 58*, by National Academy of Sciences (U.S.). Washington, D.C.: National Academy Press.

Wilson, Joseph F., and Robert L. Rausch. 1980. "Alveolar Hydatid Disease: A Review of Clinical Features of 33 Indigenous Cases of *Echinococcus multilocularis* Infection in Alaskan Eskimos." *American Journal of Tropical Medicine and Hygiene* 29 (6): 1340–55. https://doi.org/10.4269
/ajtmh.1980.29.1340.

"Wyoming Species Account: Wyoming Pocket Gopher—*Thomomys clusius*." 2020. Wyoming Fish and Game Department.

Yahalomi, Dayana, Stephen D. Atkinson, Moran Neuhof, E. Sally Chang, Hervé Philippe, Paulyn Cartwright, Jerri L. Bartholomew, and Dorothée Huchon. 2020. "A Cnidarian Parasite of Salmon (Myxozoa: *Henneguya*) Lacks a Mitochondrial Genome." *Proceedings of the National Academy of Sciences* 117 (10): 5358–63. https://doi.org/10.1073/pnas.1909907117.

Yamaguti, Satyu. 1953. *Systema Helminthum: The Digenetic Trematodes of Vertebrates*. Interscience Publishers.

Yaméogo, Laurent, Vincent H. Resh, and David H. Molyneux. 2004. "Control of River Blindness in West Africa: Case History of Biodiversity in a Disease Control Program." *EcoHealth* 1 (2): 172–83. https://doi.org/10.1007/s10393-004-0016-7.

Yanagida, Tetsuya, Jean-François Carod, Yasuhito Sako, Minoru Nakao, Eric P. Hoberg, and Akira Ito. 2014. "Genetics of the Pig Tapeworm in Madagascar Reveal a History of Human Dispersal and Colonization." *PLOS ONE* 9 (10): e109002. https://doi.org/10.1371/journal.pone
.0109002.

Yeh, Hui-Yuan, Xiaoya Zhan, and Wuyun Qi. "A Comparison of Ancient Parasites as Seen from Archeological Contexts and Early Medical Texts in China." *International Journal of Paleopathology* 25: 30–38. https://doi.org/10.1016/j.ijpp.2019.03.004.

Yensen, Eric, Teresa Tarifa, and Sydney Anderson. 1994. "New Distributional Records of Some Bolivian Mammals." *Mammalia* 58 (3): 405–14.

Yong, Ed. 2015. "How to Cure the Diseases that Nobel-Winning Drugs Cannot." *The Atlantic*, October 7, 2015. https://www.theatlantic.com/science/archive/2015/10/ivermectin-nobel-drugs -elephantiasis-filariasis-nematodes-wolbachia/409306/.

Zarlenga, Dante S., Eric P. Hoberg, and Jillian T. Detwiler. 2014. "Diversity and History as Drivers of Helminth Systematics and Biology." In *Helminth Infections and Their Impact on Global Public Health*, edited by Fabrizio Bruschi, 1–28. Vienna: Springer. https://doi.org/10.1007/978-3 -7091-1782-8_1.

Zarlenga, Dante S., Eric P. Hoberg, Benjamin Rosenthal, Simonetta Mattiucci, and Giuseppe Nascetti. 2014. "Anthropogenics: Human Influence on Global and Genetic Homogenization of Parasite Populations." *Journal of Parasitology* 100 (6): 756–72. https://doi.org/10.1645/14-622.1.

Zarlenga, Dante S., Benjamin M. Rosenthal, Giuseppe La Rosa, Edoardo Pozio, and Eric P. Hoberg. 2006. "Post-Miocene Expansion, Colonization, and Host Switching Drove Speciation among Extant Nematodes of the Archaic Genus *Trichinella*." *Proceedings of the National Academy of Sciences* 103 (19): 7354–59. https://doi.org/10.1073/pnas.0602466103.

Zarowiecki, Magdalena, and Matt Berriman. 2015. "What Helminth Genomes Have Taught Us about Parasite Evolution." *Parasitology* 142 (S1): S85–97. https://doi.org/10.1017/S00311820 14001449.

Zohar, Sandra, and John C. Holmes. 1998. "Pairing Success of Male *Gammarus lacustris* Infected by Two Acanthocephalans: A Comparative Study." *Behavioral Ecology* 9 (2): 206–11. https:// doi.org/10.1093/beheco/9.2.206.

Zoni, Ana Clara, Laura Catalá, and Steven K. Ault. 2016. "Schistosomiasis Prevalence and Intensity of Infection in Latin America and the Caribbean Countries, 1942–2014: A Systematic Review in the Context of a Regional Elimination Goal." *PLOS Neglected Tropical Diseases* 10 (3): e0004493. https://doi.org/10.1371/journal.pntd.0004493.

Index

Acanthocephala. *See* thorny-headed worm
African river blindness (onchocerciasis), 8, 19, 24, 117. *See also* blackfly; *Onchocerca volvulus, Wolbachia*
African sleeping sickness, 30. See also *Trypanosoma brucei*
agouti (*Dasyprocta*), 107, 130, 139
alcid, 51–52. *See also* puffin
Amazigo, Uche Veronica, 24
amphipod, 38, 59, 64–65, 122, 132, 139. See also *Gammarus lacustris*; *Hyalella azteca*
Ancylostoma duodenale (human hookworm), 6, 121
Anderson, Sydney, 107
Anisakis brevispiculata, 61
Anisakis simplex (herring worm), 60, 121
Apodemus uralensis (Ural field mouse), 77, *plate 16*
archaea, 114, 140
Arctic Health Research Center, 84, 113
armadillo. *See* glyptodont; pink fairy armadillo
Ascaris, 6, 10–14, 16–17, 117, 122
avermectin, 22, 140. *See also* ivermectin

Bactrian camel, 75, *plate 12*
bat, 88–89, 94, 104, 127, 129
beef tapeworm (*Taenia saginata*), 52–54, 134
Bejarano, Gaston, 108

Bering Sea, 51–52, 84; Bering land bridge, 3, 6, 52; Bering Strait, 85, 140; Chukotka Peninsula, 84; map, 85
Biomphalaria, 43, 134, *plate 7*
blackfly (*Simulium*), 8, 21, 25, 129, 150
blood fluke (*Schistosoma*), 6, 8, 39–43, 48, 117, 134, 150, *plate 6*
Bolivia, 107–109, 111–112, *plate 27, plate 28, plate 29*; Lake Titicaca, 112; map, 108. *See also* tuco-tuco
bot fly (*Gasterophilus*), 33
Boucot, Arthur, 11
Brazil, 7–8, 16, 25. *See also* Yanomami
broad fish tapeworm (*Diphyllobothrium*), *plate 2, plate 3*
brood parasite, 34
Brooks, Daniel R., 113–114

Caenorhabditis elegans, 17
California horn snail (*Cerithideopsis californica*), 43–44, 125
Campbell, William C., 22
Cardozo, Armando, 108
castration, 43, 141
Centers for Disease Control and Prevention (CDC), 73, 141
cercaria, 37, 39–40, 44, 67, 122–123, 125, 127, 134, 137, 141
Cerithideopsis californica (California horn snail), 43–44, 125

cestode (tapeworm), 34, 46–55, 56–59, 80–83, 86–89, 91–93, 97–98, 100, 111, 113, 141, *plate 2, plate 3, plate 15, plate 18, plate 19, plate 20*. See also *Diphyllobothrium; Echinococcus multilocularis; Hymenolepis diminuta; Hymenolepis lasionycteridis; Hymenolepis robertrauschi; Hymenolepis tualatinensis; Raillietina; Taenia hydatigena; Taenia krepkogorski; Taenia saginata; Taenia solium; Tetragonoporus calyptocephalus*
Chagas disease, 30, 103–105, 136, 142. See also *Trypanosoma cruzi*
Chinchorro, 104
Christmas Bird Count, 116, 142
cichlid (Cichlidae), 18, 142
citizen scientist, 116, 142
coccidia, 80, 142. See also *Eimeria*
Coitocaecum parvum, 37, 122
colonialism, 9, 19
commensal, xvii–xviii, 23, 118, 142, 150
common black spot (*Uvulifer ambloplitis*), 67, 137
common bully (*Gobiomorphus cotidianus*), 38, 122
Congo Basin, 18–19, 21, 24, 140, 146
coprolite, 4, 11
corpse lily (*Rafflesia*), 31, 133
cospeciation, 49, 51, 55